High School Math Made Simple
Third Edition

Published by TutaPoint, LLC, publishers Ryan Duques and Michael Callaghan
www.TutaPoint.com

High School Math Made Simple, Third Edition
ISBN 978-0-9834663-9-0

Written by Kara Monroe, Ph.D.
Edition author Orsolina A. Cetta
Edition co-author and math editor David Buller
Calculus chapter math editor Clayton Hall
Trigonometry chapter math editor Kara Goff
Edited by Kathleen Bergman, Marisa Nadolny, and Margaret Mary Wilson, Ed.D.
Cover Design by Dreamscape Design Group, Essex, Connecticut

TutaPoint, LLC
845 Third Avenue, 6th Floor
New York, NY 10022 USA

www.TutaPoint.com

Acknowledgements

As our authors and editors prepared for this edition of "High School Math Made Simple", we wanted to try something a little different: Engage our readers *before* we print the book. We reached out to math teachers in Maryland, Connecticut, New York, and New Jersey and asked if their classes would help us review the book. More than 30 classes took part during January 2011, providing countless suggestions and even finding a few errors. Our authors and editors would like to thank our participants.

Kate Dias, Illing Middle School, Manchester, Connecticut

Heather Sullivan, Jacob Fromerth, Alex Sanchez, Kobe Nielsen, Tommy Barrett, Ali Taseer, Olivia Rood, Ken Lippo, Marlee Coleman, Melissa Hill, Eric Weeks, Bojan Rezai, Robert Adams-Micahud, Marina Hassmann, Jessica Mhatre, Chris Choquette, TJ Nesbit, Jake Garlitz, Anna Maloney, Nouran Ghonaim, Andrew Stutts, David Stiehl, Alexis Redfield, Patrick Mockler, Joey Ruggiero, Sara Karr, Ajeet Sandhu, Evan Morrison, Kaitlin Maloney, Cameron Dias, Khylie Hill, David Nelson, Quentin Morrison, Brandon Fox, Kristen DeCiantis, Jaqueline Burns

Letitia K. Bartlett, Westminster High School, Westminster, Maryland

Tyler Bruchey, Brandon Crone, Daniel Dodrer, Dylan Congleton-Prechtel, Matt Sherlock, Chelsea Parinello, Philip Green, Whitney Schofield, Jordan Starnes, Tabetha Gratz, Sydney Wiseman, Rebecca Miller, Melissa Tyler, Lisa Redmond, Lea Schultz, Todd Tracey, Raven Watts, Kevin Kertiss, Michele Thompson, Aaron Frock, Patrick Kirby, Jesse Klaput

Wendy Adamczyk, Haddam Killingworth High School, Higganum, Connecticut

Christine Deldonna, Chad Golembeski, Nick Powers, Rachel Roberts, Alex Simms, Ana Tarbetsky, Nick Cerini

Mike Seeley, West Deptford High School, West Deptford, New Jersey

Caitie Goodfellow

Jeremy S. Youngdale, Gaithersburg High School, Gaithersburg, Maryland

Diana Quiroz, David Mensah, Mychal Walker

Donna Camardo, Paul VI High School, Haddonfield, New Jersey

Clayton R. Hall II, Hopkins School, New Haven, Connecticut

Mark Festa, Roger Lo, Taylor Waanders, Matthew Hodel, Elizabeth Peters, Arnesh Jajoo, Matthew Pun, Erik Jorgensen, Joseph Serino

Boyd Herforth

Leontine Ngoumnai, Prince George's County Public Schools, Upper Marlboro, Maryland

Tristona Fray, Amare Langford-Amrstrong, Sean Nforkah, Hawa Anthony, Janae Belt, Jamia Belton, Kayla Butler, Nancy Gutierrez, Shayla Davis

Craig Refugio, Ph.D., Charles Herbert Flowers High School, Springdale, Maryland

SY 2010-2011 classes of Dr Craig Refugio: Science and Technology Geometry, Regular Geometry, SAT Preparation and Academic Validation Project

Robert Martin, Huntingtown High School, Huntingtown, Maryland

Nicholas Bennett, School Without Walls, Washington, D.C.

Takhia Gaither, Academy for College & Career Exploration, Baltimore, Maryland

Tavon Betts

Pearl, Ithaca High School, Ithaca, New York

Kenneth Horwitz, Saddle River Day School, Saddle River, New Jersey

Students of the Honors Statistics class

Kara Goff, Century High School, Sykesville, Maryland

Patrick Komiske

Danielle Rivard, Ed.D. Bristol Eastern High School, Bristol, Connecticut

Kathleen Bergman, Valley Regional High School, Deep River, Connecticut

Luke Gagnon, Adam Jaynes, Nate Joyce, Jill Larsen, Caroline Madden, Chad Millen, Michelle Odekerken, Jacqueline Stevens, Julia Tackett, Whitney Lynn Wachtarz

Gamil Naem, Abraham Clark High School, Roselle, New Jersey

Courtney Ortiz, Devina Roman, Frank Randriantsalama, Frantz Audige, Jasmine Williams, Kelly Bermudez, Parnell Steide, Shevar Daley

Alexandra Brasoveanu-Tarpy, Thomas Wootton High School, Rockville, Maryland

Rebecca Lee, Nancy Yue, Christopher Wong, Alexander Zhu, Ganesh Subramaniam, Kevin Chuang, Victor Chen, Frank Zhao, Kate Osipova, Jessica Wilson, Michelle Chen, Jasmin Lee, Divya Sahajwalla, Victor Wang, Victoria Chai, Xuxinye Xu, Simon Dai, Akira Horiguchi, Roslyn Lee, Le Qi, Una Lee, Julia Gao, Earl Lee,

Handson Wu, Bilal Naved, Sean Wang, Defu Wan, Eileen Li, Devin Goodman, Chris Lin, Zach Siegel, David Zbarsky, Sherry Liu, April Kwan, Catherine Cheng, and Kevin Yan

Laura Scerbo, Chatham High School, Chatham, New Jersey

Brian Ballard, Peter Basso, Sophie Blumert, Jack Eisenreich, Pat Ennis, Paulete Ericksen, Amy Lee, Peter Marconi, Jack McGinley, Sara Rasmussen, Michael Strand, David Tobia, Kathleen Ughetta, Sean Whiteman, Andrew Wood, David Wych, Steven Bangs, Nate Bard, Henry Butash, James Damato, Brendan Damodaran, Julia Danitz, Maddie Dean, Grace Anne Foca, Katie Hoffman, Nathalie Ivers, Chris McAuliffe, Conner McMahon, Maggie Nelson, Jake Neumann, Tristin Pendergrast, John Radigan, Will Smart, Bethany Stachenfeld, Patrick Stock, Dylan Tencic, Doug Trumbore, Christina Kim

High School Math Made Simple

Table of Contents

Essential Math Skills

To be successful in high school mathematics, a student must be prepared for the study of algebra, geometry, data analysis, probability, trigonometry, calculus, and other disciplines. A student who is appropriately prepared to be successful in high school mathematics will be comfortable working with numbers, performing operations on numbers, working with applications involving numbers, and analyzing basic data and studies.

The purpose of this book is to outline the basic skills middle-school students or students who may be struggling in high-school mathematics need to be successful in high-school-level math, and to demonstrate those skills in an understandable way.

This book is divided into topic areas rather than grade levels. Although this book is grounded in the standards of the National Council of Teachers of Mathematics and their grade-level focal points for the study of mathematics, I realize that a student may be weak in one area of mathematics but strong in another.

Mathematics is a fascinating subject, with a thought process grounded in the world around us. Strong mathematical and critical thinking skills, as well as the problem-solving skills you develop through the study of mathematics, will provide a solid foundation for studies in business, computer science, forensics, medicine, and many other disciplines. I wish you luck in your study of mathematics.

Numbers and Operations

Parts of a Whole: Fractions

Many students enjoy math until they see fractions or decimals. For some reason, fractions and decimals strike fear into the hearts of many students who are usually confident in math class. Fractions and decimals, as well as percents, are all ways to indicate that something is part of a whole.

Let's use the example of a cookie. Say you're going to grab a snack, but the only snacks in the house are some giant cookies. We'll use a circle to see what happens to our cookie.

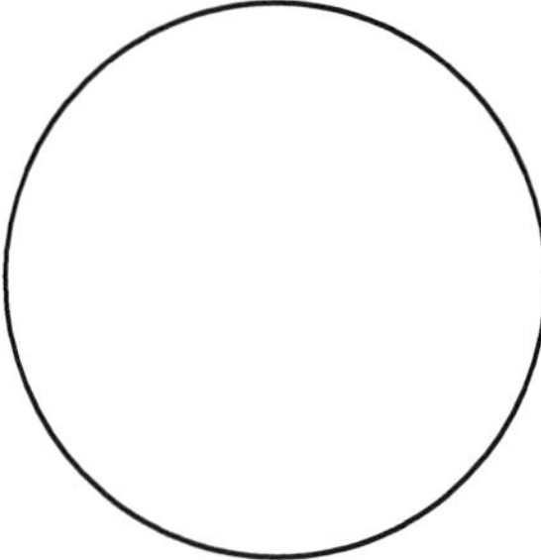

You don't want to eat a whole cookie because dinner is in just an hour. You decide to cut the cookie down the middle dividing it into two pieces that are the same size.

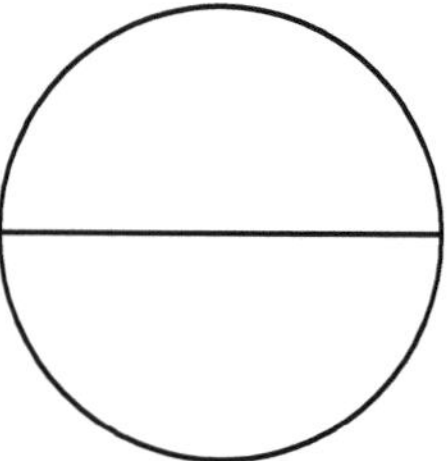

Those two pieces make up the whole cookie, but we need a way to talk about the pieces. Because we cut the cookie into two pieces, we divided one piece into two. We can talk about each of these pieces as a fraction. That fraction would look like:

$$\frac{1}{2}$$

This fraction says one whole was divided into two parts. We read this fraction as “one-half.”

What if you divided each of the halves down the middle? You would now have four pieces. If you think of what the cookie looks like now, it has been divided into four pieces.

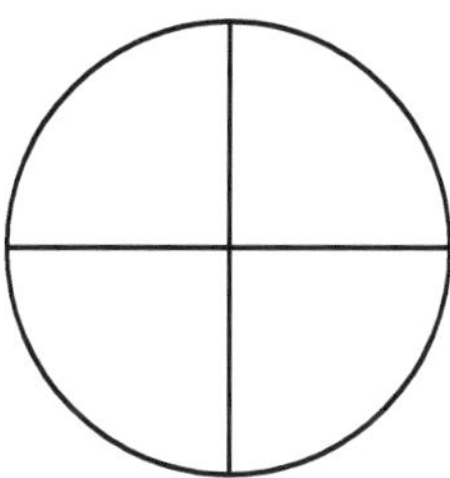

We would represent each of those four pieces as one out of four. As a fraction, we write this as:

$$\frac{1}{4}$$

We say this in one of two ways. We can either say it as “one-fourth” or we have a special way we talk about fourths – we use the term “quarters,” so we could also call it “one quarter.”

THINK: Four quarters is equal to a whole dollar. One quarter of the whole dollar is $\frac{1}{4}$.

What if we had two cookies we wanted to divide into four equal pieces? We could still state this as a fraction by placing the numeral 2 in the top of the fraction and the numeral 4 in the bottom of the fraction.

We call the top number in a fraction the **numerator** and the bottom number in a fraction the **denominator**. We would write that fraction like this:

$$\frac{2}{4}$$

If you look at this fraction, you might notice both of these numbers can be divided by 2 with no remainder. If we look at this in our picture we would see that two cookies divided into four equal pieces looks like this:

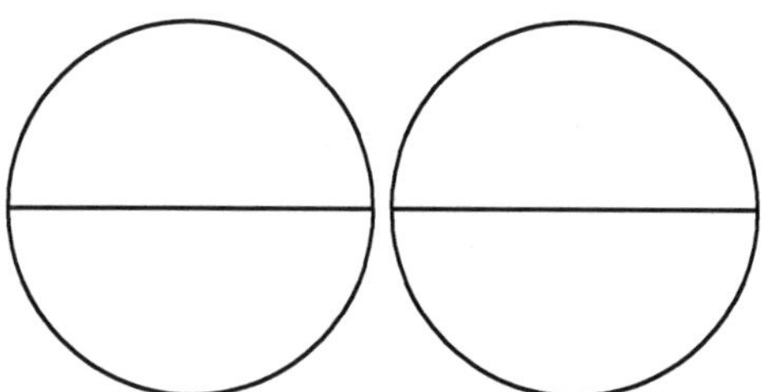

How big is each of the pieces? Each of those pieces is equal to $\frac{1}{2}$ of a cookie. We could also see this by doing what we call reducing our original fraction. When you reduce a fraction, you divide the numerator and the denominator by the same number. Let's take a look at that using our example.

$\frac{2}{4}$ → Divide both numerator and denominator by two → $\frac{1}{2}$

A fraction is completely reduced (which we call a **simplified fraction**) when the numerator and denominator can't be divided by the same whole number evenly.

When we reduce a fraction, the new fraction is equal to the first. So, you would write the problem shown above as:

$$\frac{2}{4}=\frac{1}{2}$$

TRY IT!

Write a fraction that would be equal to one piece of each of the following pictures or sets of pictures.

1. 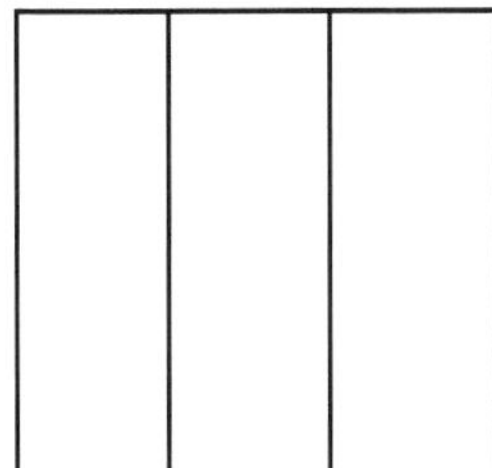

2. 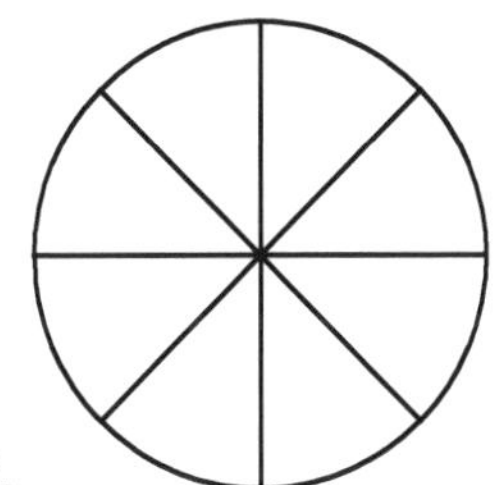

3. 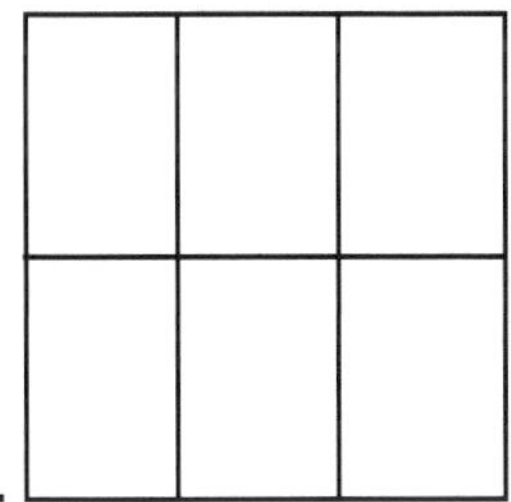

4. 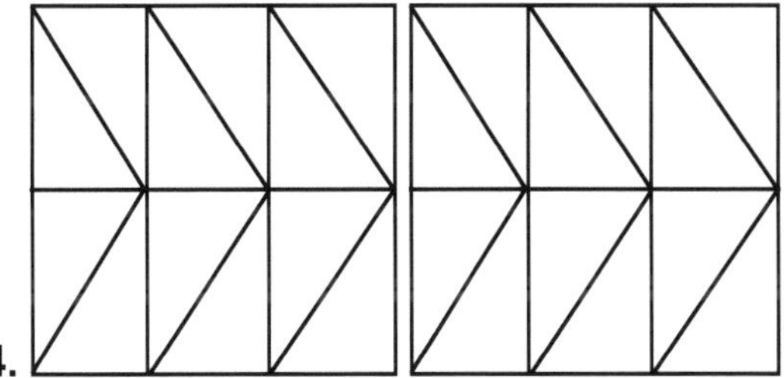

Answers

1. 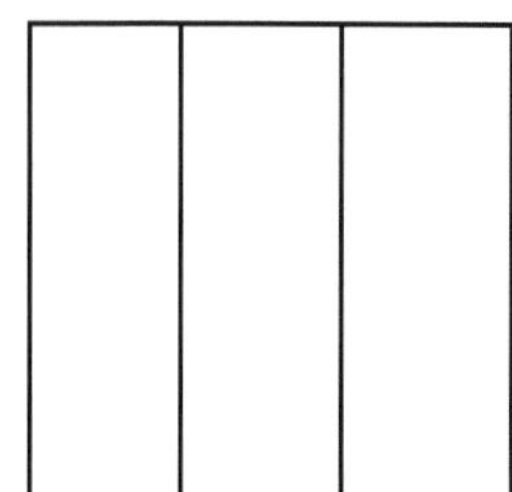

This is one square divided into three equal parts. So, one piece of this picture could be represented as $\frac{1}{3}$. This fraction already has been simplified.

2. 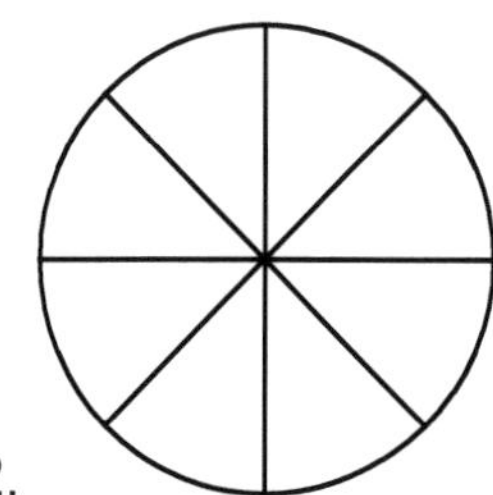

This is one circle divided into eight equal parts. So, one piece of this picture could be represented as $\frac{1}{8}$. This fraction already has been simplified.

3. 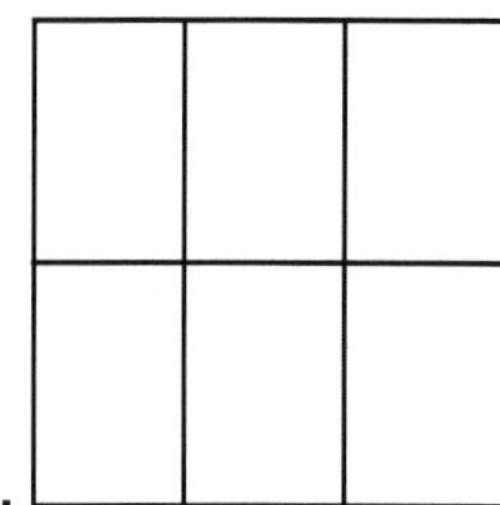

This is one square divided into six equal parts. So, one piece of this picture could be represented as $\frac{1}{6}$. This fraction already has been simplified.

4. 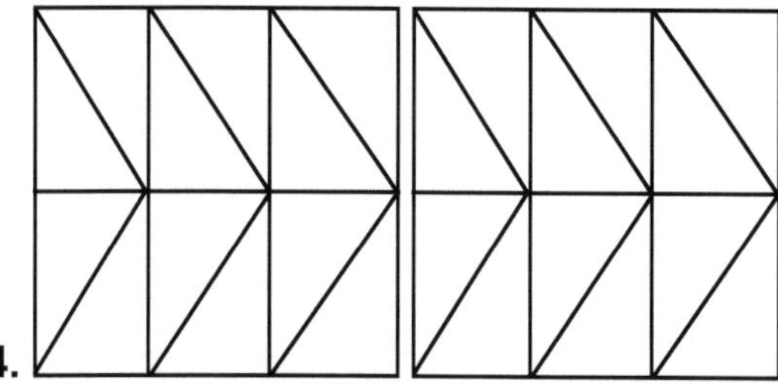

This is two squares divided into 24 equal parts. So, one piece of this picture could be represented as $\frac{2}{24}$. Both of these can be divided by 2 so the simplified form of this fraction is $\frac{1}{12}$.

TRY IT!

Practice reducing fractions by **prime factorization.** To do so, divide the numerator and the denominator of each fraction below by the same number where possible. If a fraction is already fully reduced, you do not need to do anything.

1. $\frac{14}{49}$

2. $\frac{5}{25}$

3. $\frac{3}{9}$

4. $\frac{10}{120}$

5. $\frac{4}{15}$

Answers

1. $\frac{14}{49} = \frac{\cancel{7}.2}{\cancel{7}.7} = \frac{2}{7}$

This fraction can be reduced by dividing both numerator and denominator by 7.

2. $\frac{5}{25} = \frac{\cancel{5}}{\cancel{5}.5} = \frac{1}{5}$

This fraction can be reduced by dividing both numerator and denominator by 5. Note that when we divide the numerator by 5, we end up with 1 in the numerator. In general, whenever all factors in the numerator are cancelled out, you need to leave a "1" in the numerator.

3. $\frac{3}{9} = \frac{1}{3}$

This fraction can be reduced by dividing both numerator and denominator by 3.

4. $\frac{10}{120} = \frac{1}{12}$

This fraction can be reduced by dividing both numerator and denominator by 10.

5. $\frac{4}{15} = \frac{4}{15}$

This fraction is already reduced. There is no number that divides evenly into both the numerator and the denominator.

Decimals, percents, and fractions

Decimals, percents, and fractions are a lot alike. You can rewrite decimals as percents, percents as decimals, and both as fractions. In this section we'll explore those relationships.

Let's remember a few things about decimals first. Our numbers are based on powers of 10. When you write a number like 1,385, each digit represents the number of units of each power of 10. The numeral 5 represents the number of "ones," the 8 represents the number of "tens," the 3 represents the number of "hundreds," and the 1 represents the number of "thousands." When we get smaller than a single unit (smaller than one), we use decimal numbers to represent those numbers. For example, if we want to divide a dollar into pieces we might talk about a penny – or one cent – which is written as \$0.01. If you read this as a decimal, you would read it as one-hundredth of a dollar. If we wanted to talk about a nickel, we would write this as \$0.05. This would be read as a decimal as five one-hundrodths of a dollar.

THINK: 5 cents of the total 100 cents in a dollar is \$0.05

Other decimal values can be represented using the following place values:

7	4	3	.	2	4	6	9	3
Hundreds Unit	Tens unit	Ones unit		Tenths unit	Hundredths unit	Thousandths unit	Ten-thousandths unit	Hundred-thousandths unit

If we were to read this number, we read the decimal point as the word "and." We would read this as "seven-hundred forty-three and twenty-four thousand, six hundred ninety three hundred thousandths." Notice that we only state the unit of the first and the last digits (in this case the 7 and the last 3).

If we think about the way decimals are read, it's easy to understand how they can be converted into fractions. If you read 0.37 correctly, it is read as "thirty-seven hundredths." You can easily write this as the fraction $\frac{37}{100}$.

TRY IT!

Write each of the following decimals as a fraction. Reduce the fraction to its simplest form.

1. 0.8

2. 0.29

3. 0.355

4. 2.1

5. 4.5

Answers

1. $0.8 = \frac{8}{10} = \frac{4}{5}$

This decimal is read as "eight tenths." Rewrite as a fraction and then reduce the fraction.

2. $0.29 = \frac{29}{100}$

This decimal is read as "twenty-nine one-hundredths." Rewrite as a fraction. This fraction is already in simplest form.

3. $0.355 = \frac{355}{1000} = \frac{71}{200}$

This decimal is read as "three hundred fifty-five one thousandths." Rewrite as a fraction and then reduce by dividing both numerator and denominator by 5.

4. $2.1 = 2\frac{1}{10}$

This decimal is read as "two and one tenth." Rewrite as a fraction, leaving the whole number as a whole number.

5. $4.5 = 4\frac{5}{10} = 4\frac{1}{2}$

This decimal is read as "four and five tenths." Rewrite as a fraction, leaving the whole number as a whole number. Reduce the fraction by dividing the numerator and denominator by five.

Percents can be written as both fractions and decimals. A percent can be written as a fraction if you put the percent over the number 100, because a percent is the expression of a number as a part of 100. For example, 49% can be written as $\frac{49}{100}$.

THINK: A perfect test score is 100% or entirely correct.

A percent also can be written as a decimal by moving the decimal point of the fraction two places to the left. We do this because a percent is the way of expressing the number as part of 100 or dividing by 100. You can divide any number quickly by 100 by simply moving the decimal point two places to the left.

For example, 49% can be written as the decimal 0.49. You can convert a decimal to a percent by moving the decimal point two places to the right (the same as multiplying by 100). For example, to write 0.81 as a percent, you would move the decimal point two places to the right and get 81%.

THINK: Ask yourself: How many zeros are in 100? Move two decimal places to the right. Does this pattern hold true for 1,000? Yes: move three decimal places to the right.

TRY IT!

Practice converting between fractions, decimals, and percents by filling in the table of values below with the equivalent values for each number listed.

Decimal	Fraction	Percent
0.35		
	$\frac{8}{100}$	
		72%
	$\frac{4}{5}$	
0.1		

Answers

Decimal	Fraction	Percent
0.35	$\frac{35}{100}=\frac{7}{20}$	35%
0.08	$\frac{8}{100}$	8%
0.72	$\frac{72}{100}=\frac{18}{25}$	72%
0.8	$\frac{4}{5}$	80%
0.1	$\frac{1}{10}$	10%

Comparing and Ordering Decimals, Fractions, and Percents

We also need to be able to decide between the sizes of fractions, decimals, and percents to compare them and place them in order. To do so, first convert all of the values to a common form (to a decimal, a fraction, or a percent) and then place them in order or compare.

For example, decide which is greater, 0.3 or $\frac{2}{5}$. To do this, let's convert each to a fraction. The decimal 0.3 is the same as $\frac{3}{10}$. Now, to compare the two fractions, we must convert them to the same denominator. The **least common denominator** between two fractions is found by listing the factors of each denominator and then including the greatest amount of factors of each factor for each of the denominators.

Note: Five is a prime number and does not factor.

The factors of 10 are 2 and 5.

Therefore, the numbers that we would multiply together to find our least common denominator are 2 and 5, which gives a least common denominator of 10.

So, we will convert both of these fractions to a fraction over 10.

To convert $\frac{2}{5}$ to a fraction over 10 we multiply both the numerator and the denominator by the same number. We find this number by figuring out by what number you multiply the current denominator to get the new denominator. To convert the current denominator of 5 to the new denominator of 10 we would multiply by 2. This means we multiply both numerator and denominator by 2.

$$\frac{2}{5} = \frac{4}{10}$$

Now that both fractions are written with the same denominator, we can compare them. Four tenths is greater than three tenths so $\frac{2}{5}$ is greater than 0.3. Mathematically, we would write this as

$$\frac{2}{5} > 0.3$$

Let's practice both of the skills we've just reviewed. First, let's practice finding the least common denominator and rewriting fractions with this new denominator. Then we'll practice comparing and ordering fractions, decimals, and percents.

TRY IT!

Find the least common denominator for each set of fractions listed. Then rewrite each set of fractions with the new common denominator.

1. $\frac{3}{4}$ and $\frac{2}{5}$

2. $\frac{5}{8}$ and $\frac{7}{10}$

3. $\frac{9}{20}$ and $\frac{1}{6}$

4. $\frac{2}{3}$ and $\frac{1}{5}$

5. $\frac{3}{16}$ and $\frac{7}{28}$

Answers

1. $\frac{3}{4}$ and $\frac{2}{5}$

First, we find the prime factorization of each of the current denominators. The factors of 4 are 2 and 2. The number 5 is prime. So, to find the new denominator, we multiply the largest number of each factor for each denominator. The new denominator can be found by multiplying 2, 2, and 5. This gives us a new denominator of 20.

Now we convert each fraction to a fraction over 20.

$\frac{3}{4}=\frac{15}{20}$ and $\frac{2}{5}=\frac{8}{20}$

2. $\frac{5}{8}$ and $\frac{7}{10}$

First, find the prime factorization of each denominator.

$8 = 2 \times 2 \times 2$

$10 = 2 \times 5$

To find the new denominator we multiply each factor found in its greatest quantity. Both numbers contain 2 as a factor, but the 8 includes it more times so we use all three factors of 2 as well as the factor of 5. So, the least common denominator (LCD) can be found by multiplying 2 x 2 x 2 x 5, which gives 40.

Now, we convert our fractions.

$\frac{5}{8} = \frac{25}{40}$ and $\frac{7}{10} = \frac{28}{40}$

3. $\frac{9}{20}$ and $\frac{1}{6}$

First, find the prime factorization of each denominator.

$20 = 2 \times 2 \times 5$

$6 = 2 \times 3$

The LCD can be found by multiplying 2 x 2 x 3 x 5, which gives 60. Reason: We take the prime number from the factorization in which it appears the most. If it shows once, it is part of the LCD product. In this problem, 2 shows twice in the factorization of 20, so each 2 is included in the LCD product. The 3 and 5 both show once, so they too are in the LCD product above, once each.

Now, we convert each fraction.

$\frac{9}{20} = \frac{27}{60}$ and $\frac{1}{6} = \frac{10}{60}$

4. $\frac{2}{3}$ and $\frac{1}{5}$

First, we find the prime factorization of each denominator. In this case both are prime, so we simply multiply them, which gives 15. Now, convert each fraction.

$\frac{2}{3}=\frac{10}{15}$ and $\frac{1}{5}=\frac{3}{15}$

5. $\frac{3}{16}$ and $\frac{7}{28}$

First, we find the prime factorization of each denominator.

$16 = 2\times2\times2\times2$

$28 = 2\times2\times7$

Now, we find the LCD by multiplying $2\times2\times2\times2\times7$ which is 112.

Now, convert each fraction.

$\frac{3}{16}=\frac{21}{112}$ and $\frac{7}{28}=\frac{28}{112}$

Now that we've practiced finding least common denominators and converting fractions to use the least common denominator, let's practice ordering fractions, decimals, and percents.

TRY IT!

1. Which is greater: 0.5 or $\frac{3}{4}$?

2. Which is less: 80% or 0.57?

3. Place these in order from least to greatest: $\frac{3}{8}, 0.3,$ and 41%.

Answers

1. Which is greater 0.5 or $\frac{3}{4}$?

To find which is greater, we must first write each in a common form. There are many ways to complete this. One way is to convert the decimal 0.5 to a fraction. Many people recognize 0.5 as being the same as one-half or $\frac{1}{2}$. If you didn't recognize this, that's no problem, you could still read 0.5 as five tenths and write it as a fraction then reduce it. Regardless, now we must write both $\frac{1}{2}$ and $\frac{3}{4}$ with a common denominator. The LCD is 4, so our two fractions are $\frac{2}{4}$ and $\frac{3}{4}$. Since $\frac{3}{4}$ is greater, we write the answer mathematically as

$$\frac{3}{4} > 0.5$$

2. Which is less: 80% or 0.57?

To find which is less than the other, first we must write each in a common form. There are many ways to complete this. One way is to convert both to percents. To rewrite 0.57 as a percent, we multiply it by 100 (or quickly move the decimal point two places to the right). This gives us 57%. So, 57% is less than 80%. We would write this as

$$0.57 < 80\%$$

3. Place these in order from least to greatest: $\frac{3}{8}, 0.3,$ and 41%.

We must first convert each of these values to a common form. Again, there are several approaches here, but one way is to convert each to a fraction. The decimal 0.3 will become $\frac{3}{10}$ and 41% will become $\frac{41}{100}$. Now, we must find the LCD between three numbers 8, 10, and 100. We do this in the same way we found LCDs before. We start by factoring each number:

$8 = 2 \times 2 \times 2$

$10 = 2 \times 5$

$100 = 2 \times 2 \times 5 \times 5$

We will have the numerals 2 and 5 in our LCD. We need to find the largest quantity of each factor. There are three 2's and two 5's. So, to find the LCD, we multiply:

$2\times2\times2\times5\times5$ which gives 200. Now, we write each fraction over the new LCD:

$$\frac{3}{8}=\frac{75}{200}$$

$$\frac{3}{10}=\frac{60}{200}$$

$$\frac{41}{100}=\frac{82}{200}$$

So, in order from least to greatest, the values would be $0.3, \frac{3}{8}$, and 41%.

Operations with fractions and decimals

Fractions can be added, subtracted, multiplied, and divided. To add or subtract fractions you must first find a common denominator. Once you find the common denominator, you add or subtract the numerators and write the answer over the common denominator.

For example, add $\frac{1}{5}+\frac{2}{7}$. The LCD would be 35. We rewrite the fractions as $\frac{7}{35}+\frac{10}{35}$. Now we add the new numerators and write them as a fraction over the LCD of 35. The new numerator is 17. Our answer is $\frac{17}{35}$.

We use the same process to subtract fractions. Let's subtract $\frac{3}{4}-\frac{1}{3}$.

First, find the LCD, which is 12. Now rewrite the fractions with the new LCD: $\frac{9}{12}-\frac{4}{12}$. Now we subtract the numerators and write them as a fraction over the LCD of 12. The new numerator is 5. Our answer is $\frac{5}{12}$.

To multiply fractions, you simply multiply across the numerators and across the denominators. Let's try this by multiplying $\frac{2}{3} \times \frac{4}{5}$. This gives us $\frac{8}{15}$.

Dividing fractions is done by multiplying the first fraction by the **reciprocal** of the second. The **reciprocal** is found by flipping over the fraction. Let's try this by dividing $\frac{3}{8} \div \frac{2}{3}$. To divide fractions we multiply the first fraction by the reciprocal of the second, so we will rewrite this as a multiplication problem: $\frac{3}{8} \times \frac{3}{2}$. Now, we simply multiply the numerators and the denominators. This gives us $\frac{9}{16}$.

TRY IT!

1. $\frac{1}{2} + \frac{3}{5}$

2. $\frac{7}{9} + \frac{7}{12}$

3. $\frac{1}{8} + \frac{2}{5}$

4. $\frac{3}{5} - \frac{1}{2}$

5. $\frac{4}{5} - \frac{2}{7}$

6. $\frac{7}{9} - \frac{1}{2}$

7. $\frac{3}{10} \times \frac{4}{5}$

8. $\frac{1}{8} \times \frac{7}{9}$

9. $\frac{5}{13} \div \frac{3}{5}$

10. $\frac{9}{10} \div \frac{1}{8}$

Answers

1. $\frac{1}{2} + \frac{3}{5} = \frac{5}{10} + \frac{6}{10} = \frac{11}{10}$ or $1\frac{1}{10}$

When the numerator of a fraction is larger than the denominator, the fraction is called an improper fraction. An **improper fraction** can be reduced like any other fraction by dividing a number evenly into both the numerator and denominator. An improper fraction also can be turned into a **mixed number**, $\frac{11}{10}$ or a whole number and its fractional part, by dividing the numerator into the denominator and then writing any remainder over the original denominator. In the example above, to convert to a mixed number, we divide 11 by 10. This gives 1 with a remainder of 1. So, we write the whole number (1) and then the remainder over the original denominator as ($\frac{1}{10}$).

2. $\frac{7}{9} + \frac{7}{12} = \frac{28}{36} + \frac{21}{36} = \frac{49}{36}$ or $1\frac{13}{36}$

3. $\frac{1}{8} + \frac{2}{5} = \frac{5}{40} + \frac{16}{40} = \frac{21}{40}$

4. $\frac{3}{5} - \frac{1}{2} = \frac{6}{10} - \frac{5}{10} = \frac{1}{10}$

5. $\frac{4}{5} - \frac{2}{7} = \frac{28}{35} - \frac{10}{35} = \frac{18}{35}$

6. $\frac{7}{9} - \frac{1}{2} = \frac{14}{18} - \frac{9}{18} = \frac{5}{18}$

7. $\frac{3}{10} \times \frac{4}{5} = \frac{12}{50} = \frac{6}{25}$

8. $\frac{1}{8} \times \frac{7}{9} = \frac{7}{72}$

9. $\frac{5}{13} \div \frac{3}{5} = \frac{5}{13} \times \frac{5}{3} = \frac{25}{39}$

10. $\frac{9}{10} \div \frac{1}{8} = \frac{9}{10} \times \frac{8}{1} = \frac{72}{10} = \frac{36}{5}$ or $7\frac{1}{5}$

Just like there are some rules for conducting operations with fractions, there are also rules for conducting operations with decimals. When adding or subtracting decimals, you line up the decimal points of all the numbers as well as the answer. When multiplying decimals, you count the total number of decimal places in the problem and this will be the number of decimal places in your answer. When dividing decimals the divisor — the number by which you are dividing — cannot be a decimal. To convert it and still ensure the answer is equivalent, you will move the decimal place in the divisor as well as the dividend — the number you are dividing — the same number of places. Let's look at a few examples.

Addition & Subtraction

When adding and subtracting decimals. it is important to line up the decimal points in each of the numbers as well as in the solution. This will ensure the decimal is properly placed. Once you align the decimal points, addition and subtraction occur like normal.

37.8	121.92	19.7	429.875
+ 21.9	+ 14.7	- 7.8	- 388.19
59.7	136.62	11.9	41.685

Multiplication and Division

When multiplying decimals you do not need to line up the decimal points. You simply multiply the numbers like normal and then count the number of decimal points in the original problem and place that many decimal places in the final answer.

$$\begin{array}{r} 2.3 \\ \underline{\times \quad 3} \\ 6.9 \end{array} \qquad \begin{array}{r} 3.57 \\ \underline{\times \ \ 2} \\ 7.14 \end{array} \qquad \begin{array}{r} 35.62 \\ \underline{\times\ 5.7} \\ 24934 \\ \underline{+\ 178100} \\ 203.034 \end{array} \qquad \begin{array}{r} 21.3 \\ \underline{\times\ 18.4} \\ 852 \\ 17040 \\ \underline{+\ 21300} \\ 391.92 \end{array}$$

When dividing decimals you must convert the divisor to a whole number by moving the decimal point all the way to the right. You will move the decimal point the same number of places to the right in the dividend as you move it in the divisor.

$1.2\,\overline{)12.64}$ becomes

$$\begin{array}{r} 10.4 \\ 12\,\overline{)126.4} \\ \underline{12} \\ 6 \\ \underline{0} \\ 64 \\ \underline{64} \\ 0 \end{array}$$

$3.75\,\overline{)30.375}$ becomes

$$\begin{array}{r} 8.1 \\ 375\,\overline{)3037.5} \\ \underline{3000} \\ 375 \\ \underline{375} \\ 0 \end{array}$$

TRY IT!

1. 3.8 + 9.1

2. 4.72 + 8.3

3. 9.6 + 1.75
4. 9.1 – 3.8
5. 8.3 – 4.72
6. 9.6 – 1.75
7. 3.8 x 9.1
8. 4.72 x 8.3
9. 1.75 x 9.6
10. 643.2 ÷ 6.7
11. 8.36 ÷ 1.9
12. 9.447 ÷ 6.7

Answers

1. 3.8 + 9.1 = 12.9
2. 4.72 + 8.3 = 13.02
3. 9.6 + 1.75 = 11.35
4. 9.1 – 3.8 = 5.3
5. 8.3 – 4.72 = 3.58
6. 9.6 – 1.75 = 7.85
7. 3.8 x 9.1 = 34.58
8. 4.72 x 8.3 = 39.176
9. 1.75 x 9.6 = 16.8
10. 643.2 ÷ 6.7 = 96
11. 8.36 ÷ 1.9 = 4.4
12. 9.447 ÷ 6.7 = 1.41

Rational vs. Irrational Numbers

Before we move on, it is important to distinguish between rational and irrational numbers. A **rational number** is any number that when converted to a decimal is a terminating or repeating decimal. **Irrational numbers** do not fit that list and instead are non-repeating and non-terminating. One example of an infamous irrational number is π.

Working with Integers

In addition to working with fractions, decimals, and percents, to be successful with high school mathematics, students also must be comfortable working with integers — in other words both positive and negative whole numbers. Again, there are rules we use for performing operations on integers.

Let's start with addition of integers. When adding integers with the same sign (both positive or both negative), you add the two numbers and keep the same sign. When adding integers with different signs (one positive and one negative) you subtract the two numbers and give the answer the sign of the larger number. Let's look at a few examples.

3 + 5

This is just a simple addition problem. Both numbers have the same sign (both are positive), so we add the two numbers and leave the answer positive. This gives us an answer of 8.

-2 + -9

Since these two numbers have the same sign, we will still add the numbers, but in this case since both are negative our answer will be negative. This gives us -11.

-17 + 8

In this case, the two numbers have different signs, so we will subtract the two numbers and then give the answer the sign of the larger. Since 17 is the larger of the two numbers and it is negative, our answer will be negative. This gives us an answer of -9.

4 + -3

Again, the two numbers have different signs, so, we subtract. This time the larger number is positive so our answer will be a positive number. The answer in this case is 1.

TRY IT!

1. 3 + 8
2. 15 + 9
3. 21 + 3
4. -4 + -7
5. -19 + -12
6. -1 + -2
7. 3 + -5
8. 18 + -12
9. 31 + -43
10. -19 + 27
11. -2 + 1
12. -29 + 6

Answers

1. 3 + 8 = 11
2. 15 + 9 = 24
3. 21 + 3 = 24
4. -4 + -7 = -11
5. -19 + -12 = -31
6. -1 + -2 = -3

7. 3 + -5 = -2

8. 18 + -12 = 6

9. 31 + -43 = -12

10. -19 + 27 = 8

11. -2 + 1 = -1

12. -29 + 6 = -23

When subtracting integers we are introduced for the first time to the idea that subtraction is the same as adding the opposite. Thinking of a subtraction problem in this way allows us to quickly convert the subtraction problem to an addition problem and then use the rules above to quickly finish each problem Let's look at a few examples.

5 – 2 = 5 + -2 = 3

As you already know, 5 – 2 is 3. Now, applying our rules for subtracting and then adding integers we can rewrite this as adding -2 and then use the rules for adding integers to get the same result.

3 - -5 = 3 + 5 = 8

Again, we rewrite the subtraction problem as an addition problem where we add the opposite.

Let's look at just a couple of additional examples.

-10 – 8 = -10 + -8 = -18

-8 - -6 = -8 + 6 = -2

TRY IT!

1. 18 – 7

2. 13 – 4

3. 9 – 1

4. 4 – -10
5. 16 – -17
6. 22 – -8
7. -4 – 12
8. -1 – 1
9. -49 – 22
10. -81 – -13
11. -40 – -22
12. -6 – -8

Answers

1. 18 – 7 = 11
2. 13 – 4 = 9
3. 9 – 1 = 8
4. 4 – -10 = 14
5. 16 – -17 = 33
6. 22 – -8 = 30
7. -4 – 12 = -16
8. -1 – 1 = -2
9. -49 – 22 = -71
10. -81 – -13 = -68
11. -40 – -22 = -18
12. -6 – -8 = 2

Now, we can turn our attention to the rules for multiplying and dividing integers. The rules for multiplication and division are the same, so we can go through them together. When multiplying or dividing two integers with the same sign, the answer always will be positive. When multiplying or dividing two integers with different signs (one positive and one negative), the answer always will be negative.

Let's look at a few examples.

If we want to multiply 2 x 3 you know the answer to this is 6. The answer to -2 x -3 is also 6. If we were to multiply 2 x -3 the answer would be -6. The same is true if we multiply -2 x 3; the answer is still -6.

The same rules apply to division. If we divide 8 ÷ 4 you know the answer to this is 2. The answer to -8 ÷ -4 is also 2. When dividing numbers with different signs, the answer will be negative. So:

-8 ÷ 4 = -2

And

8 ÷ -4 = 2

Let's try a few examples.

TRY IT!

1. 3 x 6
2. 9 ÷ 3
3. 5 x -3
4. 12 ÷ -2
5. -4 x -5
6. -25 ÷ -5
7. -22 x 2

8. -100 ÷ 5

9. 13 x -3

10. 49 ÷ -7

11. -3 x 6

12. 9 ÷ -3

Answers

1. 3 x 6 = 18

2. 9 ÷ 3 = 3

3. 5 x -3 = -15

4. 12 ÷ -2 = -6

5. -4 x -5 = 20

6. -25 ÷ -5 = 5

7. -22 x 2 = -44

8. -100 ÷ 5 = -20

9. 13 x -3 = -39

10. 49 ÷ -7 = -7

11. -3 x 6 = -18

12. 9 ÷-3 = -3

Evaluating Exponents

Sometimes, we need to multiply a number by itself several times, such as finding:

2 x 2 = 4

2 x 2 x 2 = 8

2 x 2 x 2 x 2 = 16

2 x 2 x 2 x 2 x 2 = 32

As you can see, writing these problems out can get quite long and it is easy to lose track of a number. To simplify our writing, we use exponents to show when we are multiplying a number by itself several times. We also call the exponents "powers." The exponent indicates the number of multiples of the number. So, if we look at the examples above, we could rewrite these as:

2 x 2 = 2^2 = 4

2 x 2 x 2 = 2^3 = 8

2 x 2 x 2 x 2 = 2^4 = 16

2 x 2 x 2 x 2 x 2 = 2^5 = 32

TRY IT!

1. Evaluate 3^4.
2. Evaluate 4^3.
3. Evaluate 10^5.
4. Evaluate 5^6.
5. Evaluate 8^3.
6. Express 5 x 5 x 5 using an exponent.
7. Express 7 x 7 x 7 x 7 x 7 x 7 using an exponent.
8. Express 6 x 6 x 6 x 6 using an exponent.
9. Express 9 x 9 x 9 x 9 x 9 x 9 x 9 x 9 x 9 x 9 using an exponent.
10. Express 1 x 1 x 1 x 1 using an exponent. Evaluate the answer.

Answers

1. Evaluate 3^4. 3^4 = 3 x 3 x 3 x 3 = 81

2. Evaluate 4^3. $4^3 = 4 \times 4 \times 4 = 64$
3. Evaluate 10^5. $10^5 = 10 \times 10 \times 10 \times 10 \times 10 = 100{,}000$
4. Evaluate 5^6. $5^6 = 5 \times 5 \times 5 \times 5 \times 5 \times 5 = 15{,}625$
5. Evaluate 8^3. $8^3 = 8 \times 8 \times 8 = 512$
6. Express 5 x 5 x 5 using an exponent.

   $5 \times 5 \times 5 = 5^3$
7. Express 7 x 7 x 7 x 7 x 7 x 7 using an exponent.

   $7 \times 7 \times 7 \times 7 \times 7 \times 7 = 7^6$
8. Express 6 x 6 x 6 x 6 using an exponent.

   $6 \times 6 \times 6 \times 6 = 6^4$
9. Express 9 x 9 x 9 x 9 x 9 x 9 x 9 x 9 x 9 x 9 using an exponent.

   $9 \times 9 \times 9 \times 9 \times 9 \times 9 \times 9 \times 9 \times 9 \times 9 = 9^{10}$
10. Express 1 x 1 x 1 x 1 using an exponent. Evaluate the answer.

    $1 \times 1 \times 1 \times 1 = 1^4 = 1$

Any number raised to the zero power (0) always will give an answer of 1. For example:

$3^0 = 1$

$5^0 = 1$

$121^0 = 1$

No matter what number you raise to the zero power, the answer always will be 1.

Number fact: If you have both positive and negative integers in the string of numbers being multiplied, an odd number of negative signs would result in a negative product. An even number of negative signs results in a positive.

For instance, $-2 \times -2 \times 2 \times -2 = -16$ and $-2 \times 2 \times 2 \times -2 = 16$.

Order of Operations

Many math problems require you to complete a series of calculations in the problem. Mathematicians use a specific order when completing the operations in a problem so the answers work out properly. This specific order is called the Order of Operations.

The Order of Operations says you will work inside the **Parentheses** first, then you complete the **Exponents.** Next you do **Multiplication and Division** in order from left to right. Finally you do **Addition and Subtraction** in order from left to right. Many students remember the order of operations by making a word out of the first letter of each of the operations (for example, PEMDAS). Other students remember the order of operations by making a sentence where the first letter of each word in the sentence is the first letter of one of the words above. A popular sentence for remembering the order of operations is **P**lease **E**xcuse **M**y **D**ear **A**unt **S**ally. Let's look at a few examples to see the Order of Operations in action.

THINK: Create your own phrase to help you remember the order of operations, PEMDAS.

Evaluate: $3^2 + 8 \times 2$

$= 9 + 8 \times 2$	Order of Operations says we evaluate the exponents first ($3^2 = 9$).
$= 9 + 16$	The next step in the Order of Operations is multiplication ($8 \times 2 = 16$).
$= 25$	Finally, we add ($9 + 16 = 25$).

So $3^2 + 8 \times 2 = 25$.

Evaluate: $(9 \div 3) - 12 \div 4$

$= 3 - 12 \div 4$	Order of Operations says we evaluate the parentheses first ($9 \div 3 = 3$).
$= 3 - 3$	Now we perform division ($12 \div 4 = 3$).
$= 0$	Finally we subtract.

So $(9 \div 3) - 12 \div 4 = 0$

Evaluate: $5^2 - (3 \times 3) + (6 - 2)$

$= 5^2 - (9) + (4)$	First we evaluate each set of parentheses ($3\times3=9$ and $6-2=4$).
$= 25 - 9 + 4$	Now we evaluate the exponent ($5^2=25$).
$= 16 + 4 = 20$	Finally, work from left to right performing addition and subtraction ($25-9=16$ and $16+4=20$).

So $5^2 - (3 \times 3) + (6 - 2) = 20$

Evaluate: $(-2)^0$

In the previous section, we learned that anything raised to the 0th power is 1, so $(-2)^0=1$. This is because we are raising the entire expression (-2) to the 0th power. However, you should be careful not to confuse this with the expression: -2^0. In this case, we are raising 2 to the 0th power, then applying the negative sign to that value: $-2^0=-1$.

TRY IT!

1. $20 + (12 \times 19)$
2. $18 \div 3 - 5$
3. $13 + (5 + 21 \times 24)$
4. $12 \div 2 \times (8 \div 2)$
5. $(18 \div 2) - 4$
6. $7 \times (6 \times 6 + 14) \times 5 + 6$
7. $21 - 18 - 8 \times 13$
8. $(3 + 10 \times 7) \times 14$

Answers

1. $20 + (12 \times 19) = 20 + 228 = 248$
2. $18 \div 3 - 5 = 6 - 5 = 1$
3. $13 + (5 + 21 \times 24) = 13 + (5 + 504) = 13 + 509 = 522$
4. $12 \div 2 \times (8 \div 2) = 12 \div 2 \times 4 = 6 \times 4 = 24$
5. $(18 \div 2) - 4 = 9 - 4 = 5$
6. $7 \times (6 \times 6 + 14) \times 5 + 6$

   $= 7 \times (36 + 14) \times 5 + 6$

   $= 7 \times (50) \times 5 + 6$

   $= 350 \times 5 + 6$

   $= 1750 + 6$

   $= 1756$
7. $21 - 18 - 8 \times 13 = 21 - 18 - 104 = 3 - 104 = -101$
8. $(3 + 10 \times 7) \times 14 = (3 + 70) \times 14 = 73 \times 14 = 1022$

Terms You Should Know

Denominator – The number beneath the bar in a fraction; this is the whole or total number of parts to create the whole.

Divisor – The number by which you are dividing in a division problem.

Dividend – The number being divided in a division problem.

Improper fraction – A fraction in which the numerator is larger than the denominator.

Least Common Denominator (LCD) – The smallest number that is a multiple of all **Denominators** – also called the lowest common denominator or the common denominator. The process for finding a least common denominator is the same as that for finding the least common multiple (LCM).

Least Common Multiple (LCM) – The smallest number that is a multiple of a given set of numbers. The least common multiple is found using the same process as that of the least common denominator.

Mixed Number – An improper fraction written as a whole number with its fractional component.

Numerator – The number above the bar in a fraction.

Quotient – The answer to a division problem.

Reciprocal – A value that when multiplied by a fraction yields a value of 1. The reciprocal is found by flipping a fraction over. (**Note:** 2 is the same as 2/1.This is true for all whole numbers. So, the reciprocal of 2/1 is $\frac{1}{2}$.

Reducing (a fraction) – The process of dividing the numerator and the denominator of a fraction by the same whole number in order to find a new fraction equal to the first.

Simplified Fraction – A fraction whose numerator and denominator can not be divided by the same whole number evenly.

Algebra I

Algebra—and more specifically Algebra I—is an advanced form of arithmetic in which students move from memorization of arithmetic facts to solving problems with a combination of arithmetic and logic. Consider the difference between these two problems. In arithmetic you might solve:

$$3 + 6 = ? \qquad \text{or} \qquad 9 - 3 = ?$$

These are easy problems that a student in algebra should be able to solve from memory. In algebra, the problems require you to find a missing value within the statement, not to find what goes on the "answer" side of the equal sign. In algebra, the problems above might be:

$$3 + ? = 9 \qquad \text{or} \qquad ? - 3 = 6$$

We've used question marks here to represent the number you're trying to identify, but in algebra we typically use letters, so the problems above become:

$$3 + x = 9 \qquad \text{or} \qquad y - 3 = 6$$

In algebra, we call these letters **variables**. To solve a problem in algebra means to find out what value the variable represents. There are many ways to solve a problem in algebra. In most cases an algebra problem can be solved both graphically and algebraically.

In the examples above ($3 + x = 9$ and $y - 3 = 6$), you could represent the problem on a number line—a figure that is one-dimensional. For example, you could represent $3 + x = 9$ like this:

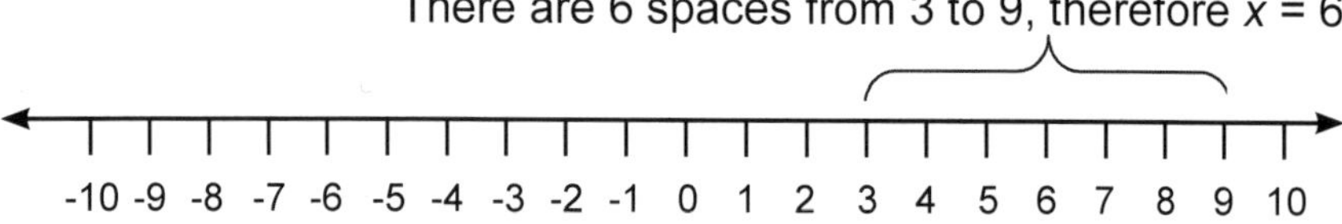

Figure 1: Using a number line to show the solution of 3 + x = 9

In algebra, you move beyond one dimension to begin representing situations and solving problems in two dimensions—on a plane. In Algebra I this means working with **linear equations,** or graphs of lines, on the plane. Algebra I students learn to represent the solution of a problem both algebraically and graphically, when appropriate.

You have used the two-dimensional plane to represent situations before when you learned to graph points like (3, 6) and (-2, 1). In algebra you move beyond graphing only those points to graphing the line that contains those points, as well as an infinite number of points that lie on the same line.

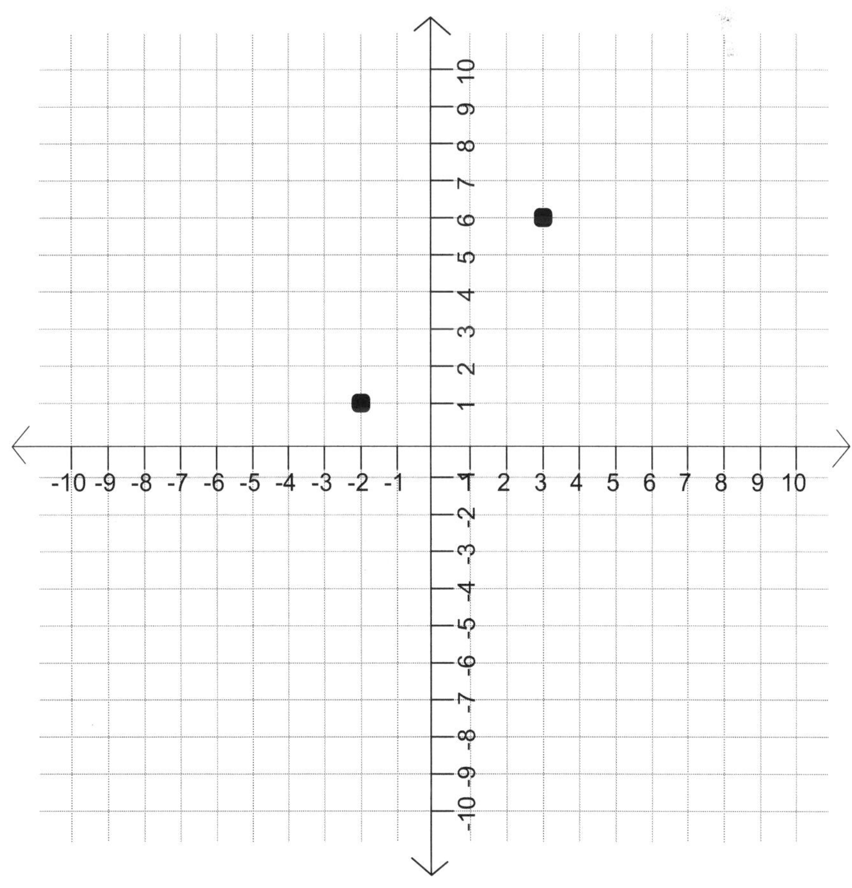

Figure 2: Graph of the points (3, 6) and (-2, 1)

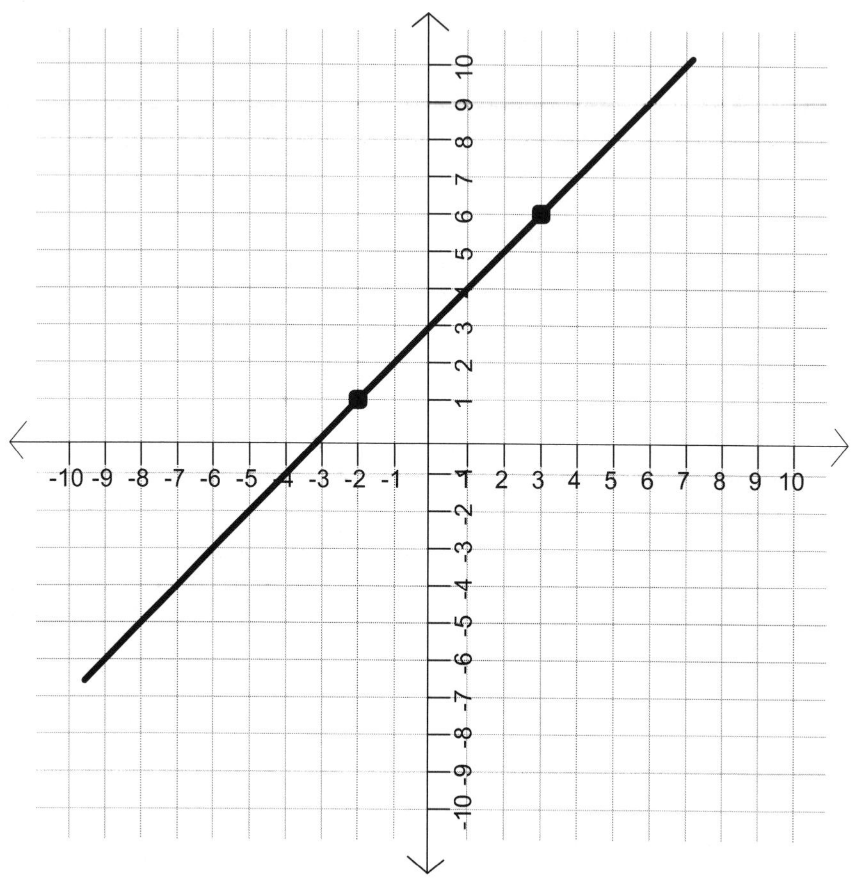

Figure 3: Graph of y = x + 3, the line that contains (3, 6) and (-2, 1)

Algebra I students learn to identify the elements of a graph, including the **slope**, the ***x*-intercept**, and the ***y*-intercept**, and they understand the importance of each of these in solving problems and representing or modeling real-world problems. We'll go over all of these terms in the problem sets below.

Finally, in Algebra I students begin to move beyond linear equations to situations that represent curves known as parabolas. The approach to these problems is primarily algebraic; however, some work with representing these problems on a graph is completed in Algebra I.

Students in Algebra I have many opportunities to work with not only the mechanics of algebra, but also with application problems that involve the use of the skills learned in algebra. Also, students have the opportunity to do additional work with the various ways to represent numbers—as fractions, decimals, and percents—and to work with ratio, proportion, and scientific notation.

Examples: Solving Equations for an Unknown Variable

Example 1: Solve 4*x* – 8 = 12.

The most important skill you learn in algebra is how to solve equations. When you complete algebra, you should be able to solve a variety of equations quickly and easily. The steps in **solving an equation**, or finding the value that the variable represents, involve getting the variable alone on one side of the equal sign. To do this, you will need to undo everything that is being done to the variable by using opposite (inverse) operations. Remember that addition and subtraction are inverses of one another as are multiplication and division.

THINK: Use Order of Operations or PEMDAS (parentheses, exponents, multiplication, division, addition, subtraction) backwards to undo or solve for an unknown value.

Let's take a look at how this works using this example:

Solve: $4x - 8 = 12$.

We want to get *x* alone on one side of the equals sign, so we need to undo what is being done to *x*. Right now, 8 is being subtracted from *x*, and 4 is being multiplied by *x* so we need to undo both of these operations by using their inverses. We begin by adding 8 to both sides.

$$\begin{array}{l} 4x - 8 = 12 \\ \quad\;\; +8 \;\; +8 \\ \hline 4x \qquad = 20 \end{array}$$

Now we are ready to divide both sides by 4. Remember that multiplication and division are inverses, so use division to "undo" multiplication.

$$\frac{4x}{4} = \frac{20}{4}$$

We are ready to state our solution: $x = 5$.

Example 2: Solve 6(*y* + 9) = 27

This is very similar to the previous problem. However, we have a choice to make at the beginning of the problem: we must decide whether to distribute the 6 through the parentheses (by multiplying both *y* and 9 by 6) or to divide by 6 first. In this instance, it does not matter which process you do first. We'll demonstrate the problem both ways so you can see that you will get the same answer, regardless of which method you choose.

Method 1: Distributing 6 through the parentheses

$6(y+9) = 27$ ←Use the distributive property to multiply the 6 through the parentheses.

$6y + 54 = 27$ ←Now this looks just like the problem above, so subtract 54 from both sides.

$$\begin{array}{rcr} 6y+54 & = & 27 \\ -54 & & -54 \\ \hline 6y & = & -27 \end{array}$$

Now the only step remaining is to undo the multiplication by dividing both sides by 6.

$$\frac{6y}{6} = \frac{-27}{6}$$

$y = -\frac{27}{6} = -\frac{9}{2}$ ←You can leave your answer as $-\frac{27}{6}$, however, most math teachers prefer that you reduce the fraction to $-\frac{9}{2}$ which is its simplified form.

Now that we've seen how to work the problem this way, let's look at Method 2, which is to divide by 6 first.

Method 2: Dividing by 6 first

$$\frac{6(y+9)}{6} = \frac{27}{6}$$

$$(y+9) = \frac{9}{2}$$

←This immediately creates a fraction, which most students would prefer to avoid until the end of the problem, but if you don't mind working with the fraction this method does result in fewer steps. Now the only step remaining is to subtract 9 from both sides.

$$\begin{array}{r} y + 9 = \frac{9}{2} \\ -9 \quad -9 \\ \hline y = -\frac{9}{2} \end{array}$$

←Most students in Algebra will use a calculator to do this problem. Fractions work easily on the calculator by using the fraction key. Check your calculator's instruction manual to learn how to use this key.

Exercises: Solving for an Unknown Value or Variable

1. Solve: $\frac{1}{2}x+8=10$.

You can begin this problem either by multiplying every term by 2 (to get rid of the fraction) or to subtract 8 from both sides. Since ½ is a simple enough fraction to work with, we'll begin by subtracting 8 from both sides to get the variable by itself.

$$\begin{array}{r} \frac{1}{2}x+8=10 \\ -8 \quad -8 \\ \hline \end{array}$$

← Subtract 8 from both sides.

$$\frac{1}{2}x=2$$

$$\frac{\frac{1}{2}}{\frac{1}{2}}x=\frac{2}{\frac{1}{2}}$$

← Divide both sides by ½ to solve for *x*.

$$x=4$$

← Simplify.

2. Solve: $5x-9=-8x+17$

There are two ways in which you could begin this problem; either by moving the variable terms to the same side of the equation or by moving the constant terms to the same side of the equation. You may also want to look ahead a bit to make the problem as easy to work with as possible. For instance, if we choose to move the variable terms first, we can either move 5*x* by subtracting 5*x* from both sides (giving a negative value on each side) or we can move -8*x* by adding 8*x* to both sides (which gives a positive value). This is a better choice because we will not have to work with a negative variable.

$$\begin{array}{l} 5x-9=-8x+17 \\ \underline{+8x \qquad\quad +8x} \\ 13x-9= \qquad\quad 17 \end{array}$$

← Add $8x$ to both sides.

$$\begin{array}{l} 13x-9=17 \\ \underline{\quad\;\; +9 \;\; +9} \\ 13x \qquad =26 \end{array}$$

← Add 9 to both sides to isolate the variable term.

$$\frac{13x}{13}=\frac{26}{13}$$

← Divide both sides by 13 to solve for *x*.

$$x=2$$

← Simplify.

3. **Solve:** You have the choice between two new plans for cellular phones. In the first plan, you pay 15 cents per minute. In the second plan, you pay a fee of $10 per month plus 5 cents per minute. For how many minutes must you talk in order for the two plans to cost the same?

To solve this problem, you need to write equations that represent each of the two plans. Since the amount that changes (the variable) is the number of minutes you talk, let's use the variable *m*. Now we can write an equation for each plan:

For the first plan, you pay 15 cents per minute, so this equation is simply 0.15*m*.

For the second plan, you pay $10, plus 5 cents per minute. For this equation you need to add together the $10 base fee and the 5 cents per minute. This equation would look like: 10 + 0.05*m*.

Now, we need to set the two equations equal and solve for *m:*

$$0.15m = 10 + 0.05m$$

To solve this equation, we can either multiply both sides of the equation by 100 to clear the decimals or we can simply solve the equation with the decimals. This is a simple enough problem that we'll just solve with the decimals. Note that, if you wanted to clear the decimals, multiplying both sides of the equation by 100 would yield the equation $15m = 1000 + 5m$. You'll get the same answer regardless of which equation you use.

To solve 0.15*m* = 10 + 0.05*m*, you need to subtract 0.05*m* from both sides to get the variable terms together on one side of the equation.

$$0.15m = 10 + 0.05m$$

$$\begin{array}{rl} 0.15m = 10 + 0.05m & \\ -0.05m \quad\quad -0.05m & \leftarrow \text{Subtract -0.05}m \text{ from both sides. Simplify.} \\ \hline 0.10m = 10 & \end{array}$$

$$\frac{0.10}{0.10}m = \frac{10}{0.10} \quad \leftarrow \text{Divide both sides by 0.10 to solve.}$$

$$m = 100 \quad \leftarrow \text{Simplify.}$$

This means if you talk for 100 minutes a month, the cost for the two plans would be equal.

4. **Solve:** You are planning a road trip and want to set a budget for gas mileage. To do this, you need to know how many gallons of gas you will need for the trip. The trip you are planning is 400 miles in length. You know that you can travel 300 miles on 13 gallons of gas. How many gallons will you need to drive 400 miles? (Round to the nearest tenth.)

This problem is easiest to solve by setting up two ratios and setting them equal to one another in a proportion. The first ratio is the amount of miles you drove in comparison to the required 13 gallons of gas. This ratio is:

$$\frac{300}{13}$$

The second ratio should be set up in the same way, with the length of the trip you are planning being compared to the number of gallons of gas you will need. Since we don't know the number of gallons of gas needed, let this value be x. This ratio is:

$$\frac{400}{x}$$

To solve the problem, set the two ratios equal to one another (this forms what is called a proportion).

$$\frac{300}{13} = \frac{400}{x}$$

To solve a proportion, cross-multiply. Cross-multiplying gives us:

$$\frac{300}{13} \times \frac{400}{x}$$

$300x = 13 \cdot 400$	← Cross-multiply.
$300x = 5200$	← Simplify.
$\frac{300}{300}x = \frac{5200}{300}$	←Divide both sides by 300 to solve for x.
$x = 17.3$	←You will need 17.3 gallons of gas for the trip.

5. **Solve:** $8(3x - 7) = 2(6x - 4)$

To begin this problem, clear the parentheses on both sides by distributing:

$$8(3x - 7) = 2(6x - 4)$$

$24x - 56 = 12x - 8$	←Multiply using the distributive property.
$\begin{array}{r} 24x - 56 = 12x - 8 \\ -12x \quad\quad -12x \\ \hline 12x - 56 = -8 \end{array}$	←Subtract $12x$ from both sides and simplify.

$$\begin{array}{l} 12x - 56 = -8 \\ \quad\;\; +56 \;\; +56 \\ \hline 12x \qquad = \;\; 48 \end{array}$$

←Add 56 to both sides to isolate *x*. Simplify.

$$\frac{12}{12}x = \frac{48}{12}$$

←Divide both sides by 12.

$x = 4$ ←Simplify.

TRY IT!

1. Solve for x: 13x + 147 = -73
2. Solve for x: – (2 x - 4) = -24
3. Solve for x: 12 – 18 = -4 (2x – 7) – 3x
4. Solve for x: -2 (3x – 12) = 10- x

Answers

1. Solve for x: $13x + 147 = -73$

To begin this problem, we want to **isolate the variable**:

$$\begin{array}{l} 13x + 147 = -73 \\ \quad\;\; -147 \;\; -147 \\ \hline 13x \qquad = -220 \end{array}$$

← Subtract 147 from both sides. Simplify.

$$\frac{13x}{13} = \frac{-220}{13}$$

←Divide both sides by 13.

$x = -16.92$ ←Simplify.

2. Solve for x: $-(2x - 4) = -24$

To begin this problem, clear the parentheses by distributing the negative sign.

$$-(2x - 4) = -24$$

$-2x+4=-24$ ←Multiply using the distributive property.

$$\begin{array}{l} -2x+4=-24 \\ \quad\;\; -4 \quad -4 \\ \hline -2x \quad\; =-28 \end{array}$$

← Subtract 4 from both sides. Simplify.

$\frac{-2x}{-2}=\frac{-28}{-2}$ ←Divide both sides by -2.

$x=14$ ←Simplify.

3. Solve for x: $12-18=-4(2x-7)-3x$

To begin this problem, clear the parentheses by distributing the negative 4 on the left side of the equals sign and by simplifying the right side of the equals sign.

$12-18=-4(2x-7)-3x$

$-6=-8x+28-3x$ ←Subtract. Multiply using the distributive property.

$-6=-11x+28$ ← Combine like terms on the right side.

$$\begin{array}{l} -6=-11x+28 \\ -28 \qquad\quad -28 \\ \hline -34=-11x \end{array}$$

←Subtract 28 from both sides. Simplify.

$\frac{-34}{-11}=\frac{-11x}{-11}$ ←Divide both sides by -11.

$x=\frac{34}{11}$ ←Simplify.

4. Solve for x: $-2(3x-12)=10-x$

To begin this problem, clear the parentheses by distributing the negative 2 on the left side of the equal sign.

$-2(3x-12)=10-x$

$-6x+24=10-x$ ← Multiply using the distributive property.

$$\begin{array}{r} -6x+24=10-x \\ +\ x \qquad\qquad +x \\ \hline -5x+24=10 \end{array}$$ ← Add x to both sides. Simplify.

$$\begin{array}{r} -5x+24=\ 10 \\ -24\ \ -24 \\ \hline -5x \qquad =-14 \end{array}$$ ←Subtract 24 from both sides. Simplify.

$$\frac{-5x}{-5}=\frac{-14}{-5}$$ ←Divide both sides by -5.

$$x=\frac{14}{5}$$ ←Simplify.

Example 3: Find the slope, *x*-intercept, and *y*-intercept of $\frac{1}{2}x + y = 3$. Graph the line.

Before we can begin this problem, we need to define the slope, *x*-intercept, and *y*-intercept of a line. The **slope** of a line is defined as the rise of the line divided by the run of the line, or, put another way, it is the change in *y* values divided by the change in *x* values. This is also known as the rate of change, or the steepness of the line.

There are two ways to find the slope of a line. The first is to use two points on the line and compute the slope using a formula. The second is to use the equation of a line to find the slope by solving the equation for *y* and looking at the coefficient of *x*—this number is always the slope. Solving for *y* gives you a format called the **slope-intercept** form of a line. This form looks like $y = mx + b$. In this form, *m* always represents the slope, and *b* is the *y*–intercept, the point at which the line intersects the *y*-axis. The second method is the method we will use here.

The next term is *x*-intercept. The ***x*-intercept** of a line is the point at which the line intersects the *x*-axis on the graph. It is also the place where *y* is equal to 0.

The final term is *y*-intercept. The ***y*-intercept** of a line is the point at which the line crosses the *y* axis on the graph. It is also the place where *x* is equal to 0.

Now that we have defined our terms, let's take a look at how we find the slope, *x*-intercept, and *y*-intercept. First, we'll solve for *y* to find the slope.

$\frac{1}{2}x + y = 3$ ← To get *y* by itself we subtract $\frac{1}{2}x$ from both sides.

$-\frac{1}{2}x \quad -\frac{1}{2}x$

$y = -\frac{1}{2}x + 3$ ← This equation is now in slope-intercept form.

Since the $-\frac{1}{2}$ is the coefficient of *x*, this is your slope, so $m = -\frac{1}{2}$.

THINK: A line that decreases as we move from left to right has a negative slope. With this slope, -½, as the line moves two units to the right, it will move down by one unit. Another way to read this slope is that the line moves up one unit for every two units we move to the left. Notice that the line must move left or down in a negative slope. Moving left or down represents negative movements on the coordinate grid; moving to the right or up represents positive movements; and moving up or down represent changes in the y-values, and moving right or left represent changes in the x-values.

Now, we'll find the *x*-intercept. This is the *x* value when *y* = 0. You can find this by simply substituting 0 in for *y* and solving for *x*.

$\frac{1}{2}x + 0 = 3$

$\frac{2}{1} \cdot \frac{1}{2}x = \frac{2}{1} \cdot 3$ ← To divide by a fraction, you multiply by its reciprocal.

$x = 6$

So, the *x*-intercept is the point (6, 0).

To find the *y*-intercept, substitute 0 in for *x*.

$$\frac{1}{2}(0)+y=3$$

$$y=3$$

So, the *y*-intercept is (0, 3).

Now we can use these two points or one of these points and the slope to graph the line:

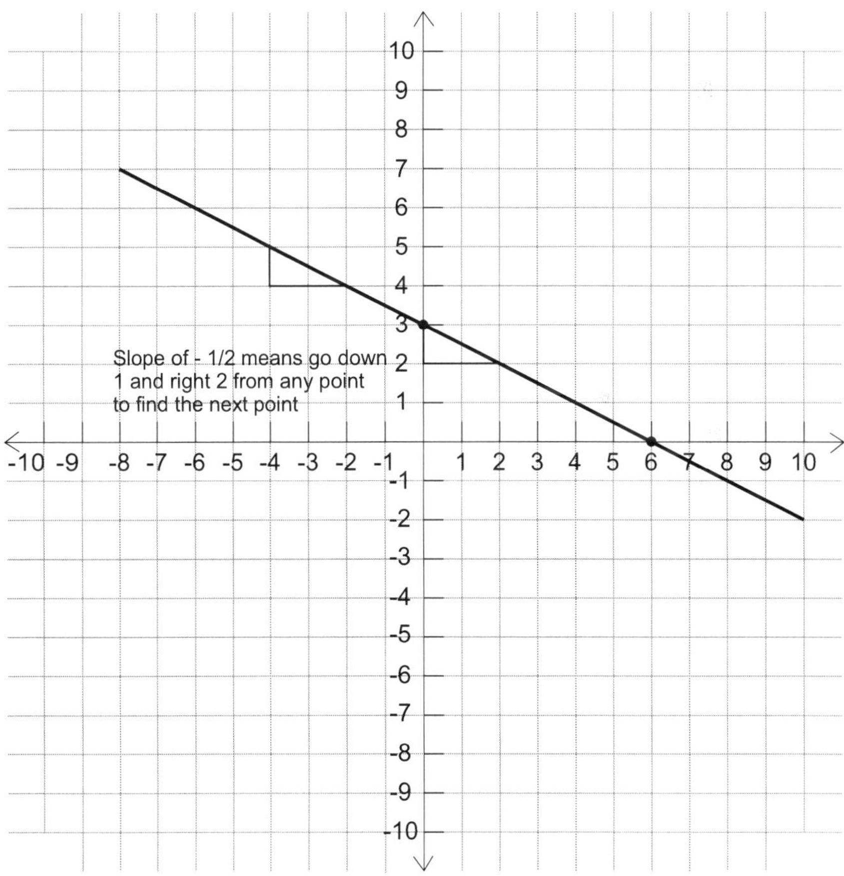

Figure 4: This shows the graph of *y* = -1/2 *x* + 3. Notice that we graphed the points of the *x*- and *y*-intercepts and have shown how the slope can be used to find any point on the line.

Exercises: Working with Linear Equations

1. Graph $3x - y = 4$ by finding at least two points on the line.

To find points on the line and graph the line, you pick a value for either x or y, then substitute that value in and use it to solve for the other value. You can pick any value you want, but it is often easy to use small numbers and 0. For this example, we'll use the numbers 1 and -1 for x. We only need to find two values to graph the line, but we'll also find the values for $x = 0$.

Generally, it is good practice to encourage students to check their solutions by plugging the value back into the original equation. You may want them to solve for y first to make it easier to calculate.

X	Y	
1	-1	$3(1) - y = 4 \rightarrow 3 - y = 4 \rightarrow y = -1$
0	-4	$3(0) - y = 4 \rightarrow 0 - y = 4 \rightarrow y = -4$
-1	-7	$3(-1) - y = 4 \rightarrow -3 - y = 4 \rightarrow y = -7$

Now plot these points on the graph and connect the points with a line.

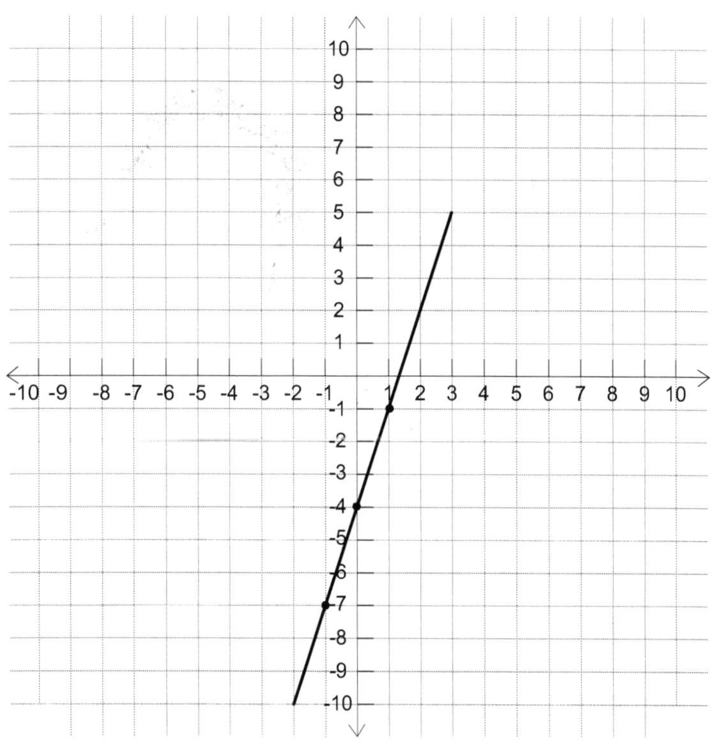

2. Graph $2x + 3y = 12$ using the slope and y-intercept.

To graph a line using the slope and the y-intercept, you must put the equation into the slope-intercept form of a line, which is $y = mx + b$, where m is the slope of the line and b is the y-intercept. You can put any equation into this form by solving for y.

$$2x + 3y = 12$$

$$\begin{array}{l} 2x + 3y = 12 \\ \underline{-2x \qquad\quad -2x} \\ 3y = -2x + 12 \end{array}$$ ← Subtract $2x$ from both sides to start getting y by itself.

$$\frac{3y}{3} = \frac{-2x}{3} + \frac{12}{3}$$ ← Divide each term by 3 to isolate y.

$$y = -\frac{2}{3}x + 4$$ ←Simplify.

Now that you have the equation in slope-intercept form, you can use the y-intercept and the slope to graph the line quickly. You know that the y-intercept (or the place where the graph crosses the y-axis) is at 4, which means it crosses at the point (0, 4). You also know that the slope is $-\frac{2}{3}$. This means that for any point on the graph, the change in y from that point to the next point is going to be –2 and the change in x from that point to the next point is going to be 3. So, you can start at the y-intercept and go down 2 and right 3 to find the next point on the line, which will be (3, 2):

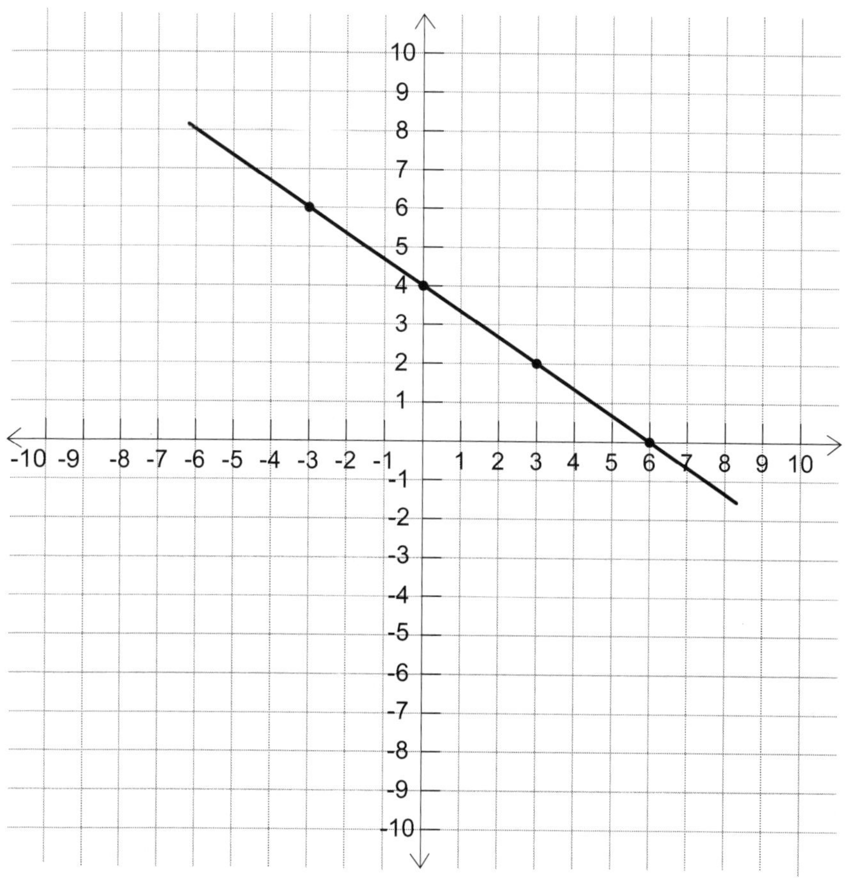

3. Find the equation of the line containing the points (-3, 8) and (7, -2).

To find the equation of a line, you will need to find the slope, as well as one of the points on the line. The slope can be found by substituting the values for two points into the formula for slope, which is:

$$m = \frac{y_2 - y_1}{x_2 - x_1}$$

When you substitute in the two points (-3, 8) and (7, -2) this gives:

$$m = \frac{-2-8}{7--3} = \frac{-10}{10} = -1$$

Now use the slope and one of the points (either point is fine; they both lead to the same equation) in the point-slope form of the line to find the equation. The point-slope form of the line looks like:

$$y - y_1 = m(x - x_1)$$

When you substitute the slope (-1) and one of the points (-3, 8) into this formula, you get:

$y - 8 = -1(x - -3)$, which we can simplify to:

$y - 8 = -1(x + 3)$

Now simplify this equation. If you solve for *y* you'll put the equation into slope-intercept form. If you move the *x* and *y* term to the same side of the equation and place the constant term on the other side of the equals sign, you will put the equation in standard form. We'll put this equation in standard form.

$y - 8 = -1(x + 3)$

$y - 8 = -x - 3$ ← Distribute -1 through the parentheses.

$$\begin{array}{l} y - 8 = -x - 3 \\ \underline{+x \qquad\quad +x} \\ x + y - 8 = -3 \end{array}$$

← Add *x* to both sides. Simplify.

$$\begin{array}{l} x + y - 8 = -3 \\ \underline{\qquad\quad +8 \quad +8} \\ x + y \qquad = 5 \end{array}$$

← Add 8 to both sides. Simplify. The equation is now in standard form.

4. Find the equation of the line parallel to the line *x* – 2*y* = 6 that passes through the point (4, 7).

A line is parallel to another line if the slopes are equal. To solve this problem, we'll need to find the slope of the line *x* – 2*y* = 6 and then use that slope and the point (4, 7) to find the equation of the new line. Let's begin by finding the slope of the line *x* – 2*y* = 6 by solving for *y*.

$x - 2y = 6$

$$\begin{array}{l} x - 2y = 6 \\ \underline{-x \qquad -x} \\ -2y = -x + 6 \end{array}$$ ← Subtract *x* from both sides. Simplify.

$$\frac{-2y}{-2} = \frac{-1x}{-2} + \frac{6}{-2}$$ ← Divide both sides by -2 to solve for *y*.

$$y = \frac{1}{2}x - 3$$ ← Simplify.

In this form, we can see just by looking that the slope of the line is $\frac{1}{2}$. Now we need to find the equation of the line through the point (4, 7) with the slope of $\frac{1}{2}$.

To do this, we'll substitute these values into $(y - y_1) = m(x - x_1)$. This gives us $(y - 7) = \frac{1}{2}(x - 4)$.

This answer is fine and you could stop here, but most instructors will ask you to put the equation in slope-intercept form or standard form, so you may need to continue working with the equation to put it into one of those forms.

TRY IT!

1. Find the *x*- and *y*-intercept of the line: $2x + 8y = 16$. Graph the line using the intercepts.
2. Graph by finding at least two points on the line: $-2x - y = 8$
3. Find the slope and y-intercept of the line: $4x - 2y = 12$. Graph the line using the slope and *y*-intercept.
4. Find the equation of the line which passes through the points (3, 5) and (6, 0).

5. Find the equation of the line parallel to $y=-7x+4$ that passes through (-1, 0).

Answers

1. Find the *x*- and *y*-intercept of the line: 2x + 8y = 16; graph the line using the intercepts.

To find the *x*-intercept — this is the *x* value when *y* = 0 —simply substitute 0 in for *y* and solve for *x*.

$2x+8(0)=16$

$2x=16$

$x=8$

So, the *x*-intercept is the point (8, 0).

To find the *y*-intercept, substitute 0 in for *x*.

$2(0)+8y=16$

$8y=16$

$y=2$

So, the *y*-intercept is (0, 2).

Now we can use these two points, or one of these points and the slope, to graph the line:

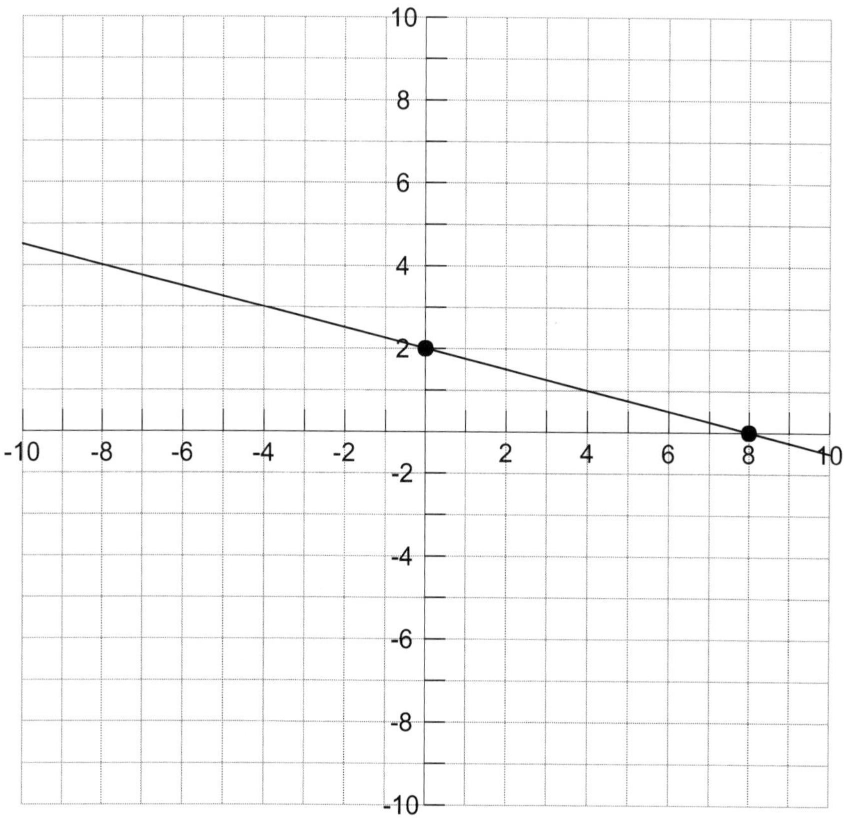

2. Graph by finding at least two points on the line: - 2x – y = 8

Remember, to find points on the line and graph the line, pick a value for either *x* or *y*, then substitute that value in and use it to solve for the other value. You can pick any value, but it is often easy to use small numbers and 0. For this example, we'll use the numbers 1 and -1 for *x*. We only need to find two values to graph the line.

$$-2(1) - y = 8$$

$$-2 - y = 8$$
$$-y = 10$$
$$y = -10$$

So, one point on the line is (1, -10).

$$-2(-1) - y = 8$$
$$2 - y = 8$$
$$-y = 6$$
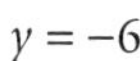
$$y = -6$$

Another point on the line is (-1, -6).

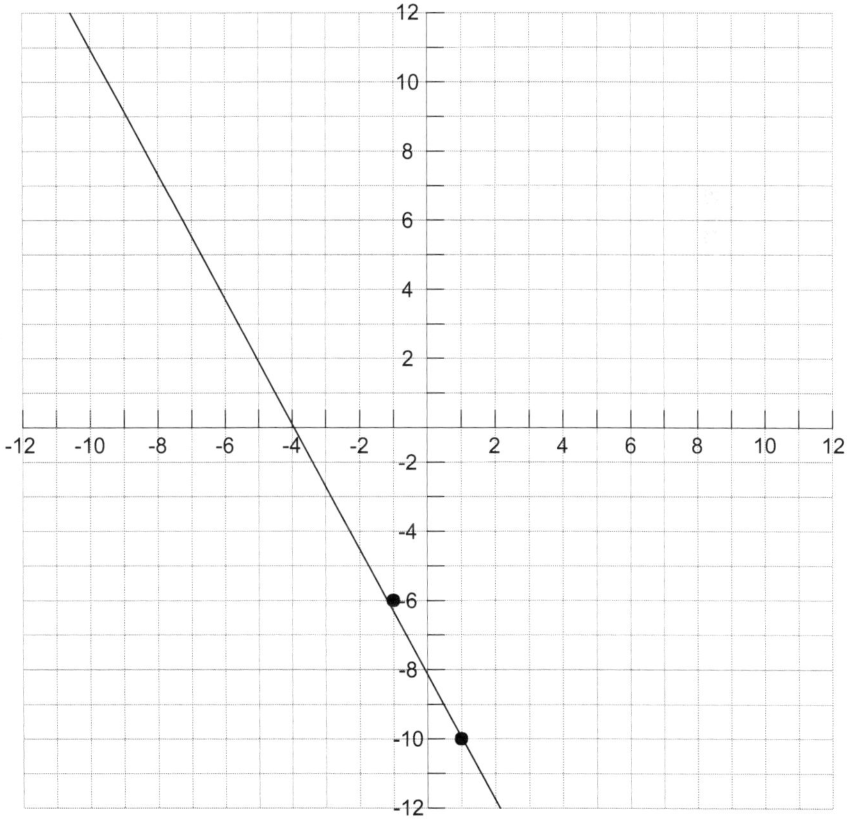

3. Find the slope and *y*-intercept of the line: 4x – 2 y = 12; Graph the line using the slope and *y*-intercept.

Remember, to find the slope and *y*-intercept of the line, we will need to manipulate the equation into $y = mx + b$ form.

$$4x - 2y = 12$$
$$-2y = 12 - 4x$$
$$y = \frac{12 - 4x}{-2}$$
$$y = -6 + 2x$$
$$y = 2x - 6$$

Once this equation is simplified, we can see that the slope, or m = 2, and the *y*-intercept is -6 or (0, -6).

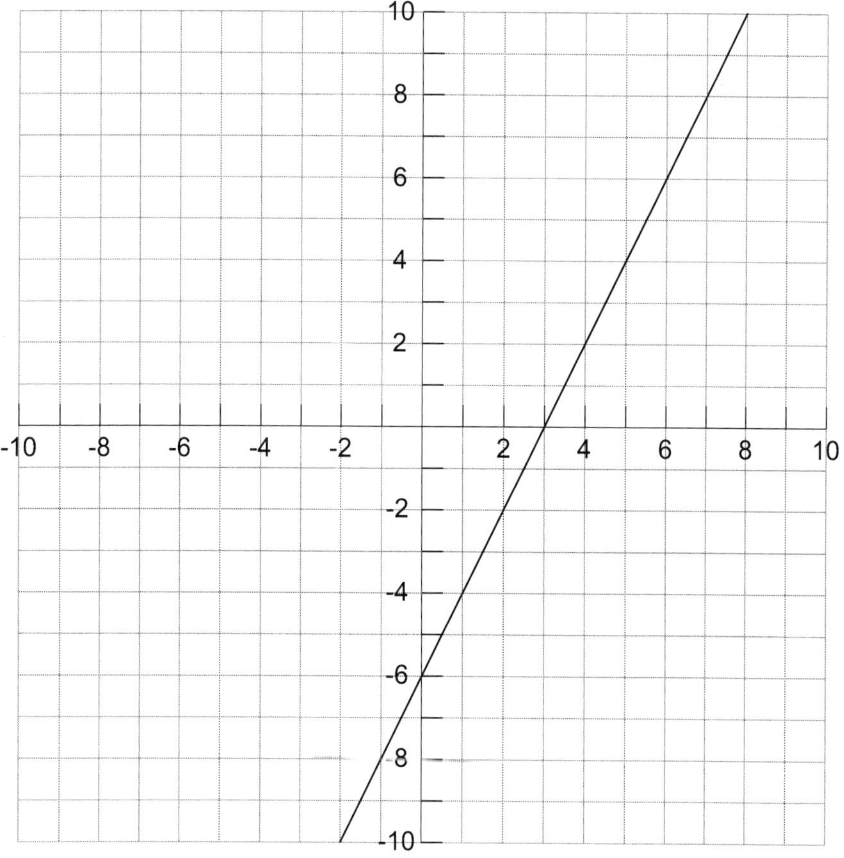

4. Find the equation of the line with the points (3, 5) and (6, 0).

Remember, to find the equation of a line, you will need to find the slope, as well as one of the points on the line. The slope can be found by substituting the values for two points into the formula for slope, which is:

$$m = \frac{y_2 - y_1}{x_2 - x_1}$$

When you substitute in the two points (3, 5) and (6, 0) this gives:

$$m = \frac{0-5}{6-3} = \frac{-5}{3}$$

Now use the slope and one of the points (either point is fine; they both lead to the same equation) in the point-slope form of the line to find the equation. The point-slope form of the line looks like:

$$y - y_1 = m(x - x_1)$$

When you substitute the slope $\left(\frac{-5}{3}\right)$ and one of the points (6, 0) into this formula, you get:

$$y - 0 = \frac{-5}{3}(x-6)$$

Now simplify this equation in whatever form is appropriate. If you solve for *y* you'll put the equation into slope-intercept form. If you move the *x* and *y* term to the same side of the equation and place the constant term on the other side of the equals sign, you will put the equation in standard form. We'll put this equation in slope-intercept form:

$$y - 0 = \frac{-5}{3}(x-6)$$

$y = \frac{-5}{3}x + \frac{30}{3}$ ←Distribute.

$y = \frac{-5}{3}x + 10$ ←Simplify.

5. Find the equation of the line parallel to y = - 7 x + 4 that passes through (-1, 0).

A line is parallel to another line if the slopes are equal. To solve this problem, we'll need to use the slope of the given line, -7 and the point (-1, 0) to find the equation of the new line.

To do this, we'll substitute these values into $(y-y_1)=m(x-x_1)$. This gives us $(y-0)=-7(x+1)$. This answer is fine and you could stop here, but most instructors will ask you to put the equation in slope-intercept form or standard form, so you may need to continue working with the equation to put it into one of those forms.

Example 4: *y* varies directly as *x*. If *y* is 8, then *x* = 16. Write an equation for this relationship.

Many situations in life can be represented in terms of direct or inverse variation. In models that use direct variation, when one quantity increases, another increases (or as one quantity decreases the other decreases). **Direct variation** equations are represented by the model *y* = *kx* where *k* is the constant of variation. A model that represents indirect variation represents a situation in which one quantity increases as the other decreases and vice versa.

Inverse variation equations are represented by the model $y=\frac{k}{x}$ where *k* is the **constant of variation**. The goal in a problem like this one is to find the value of *k* and write the equation that models the relationship. To solve this example, we substitute in 8 for *y* and 16 for *x* and then solve for *k*:

$y=kx$	← This is the equation for the model of direct variation.
$8=k.16$	← Substitute 8 for *y* and 16 for *x*.
$\frac{8}{16}=k.\frac{16}{16}$	← Divide both sides by 16 to solve for *k*.
$y=\frac{k}{x}$	← Simplify the fraction $y=kx$.
$y=\frac{1}{2}x$	← Rewrite the model for direct variation, using the value you found for *k*.

Exercises: Working with Direct Variation

1. **Solve:** The number of hours you work varies directly with the amount of your weekly salary. In a certain week, you worked 22 hours and had a salary of \$209.88. Find an equation that models this situation.

Recall that direct variation is represented by the model $y = kx$. In this example, the equation $209.88 = k(22)$. We'll use this to solve for k.

$\frac{209.88}{22} = k(\frac{22}{22})$ ← Divide both sides by 22 to solve for k.

$9.54 = k$ ← Simplify.

Now, rewrite the variation equation, substituting in the value for k:

$$y = 9.54x$$

TRY IT!

1. An automobile is traveling at the constant rate of 50 mph. The distance varies directly as time progresses. Find an equation to model this situation.

Answers

1. Recall that direct variation is represented by the model $y = kx$. In this example, the distance traveled is $d = 50t$.

Factoring Polynomial Expressions

Example 5: Factor $x^2 - 5x = 36$. Solve.

Factoring is one of the most important skills that a student will learn in Algebra I. It is critical in the ability to solve problems in upper level mathematics courses. To solve an equation, it must be set equal to 0, so the first step in this problem is to subtract 36 from both sides to get one side of the equation equal to 0.

$x^2 - 5x = 36$

$x^2 - 5x - 36 = 36 - 36$ ← Subtract 36 from both sides to set the right side equal to 0.

$x^2 - 5x - 36 = 0$

Now, you want to identify factors of the last term (-36), which, when added together, give you the middle term (-5). So, let's list the factors of -36 and add up each pair to see which one gives us -5.

-36			Sum
Factors			
-1	36	⇨	35
1	-36		-35
-2	18		16
2	-18		-16
-3	12		9
3	-12		-9
-4	9		5
4	-9		-5
-6	6		0

You can see from the table that the only factors of -36 that combine to give you -5 are 4 and -9. So, to factor the equation, we need to write two factors and set them equal to 0.

$(x+4)(x-9) = 0$

Now we set each of the separate factors equal to 0, and solve each of the resulting equations. So, either:

$x + 4 = 0$ or $x - 9 = 0$

$$\begin{array}{rcl} x + 4 &=& 0 \\ -4 && -4 \\ \hline x = -4 \end{array} \qquad \begin{array}{rcl} x - 9 &=& 0 \\ +9 && +9 \\ \hline x = 9 \end{array}$$

This problem has two solutions. They are $x = -4$ or $x = 9$.

Note: Solutions are known as zeros, roots, or *x*-intercepts. Remember that *x*-intercepts are the points where y = 0 and the line intersects the x-axis.

Exercises: Problems Involving Factoring

1. Factor: $6x^2 - 28x - 48$.

The goal of this problem is to factor it completely. There is a general strategy that you should use for factoring any expression. This strategy is:

a. Factor out the greatest common factor (GCF) if there is one.

b. Count the number of terms in the expression and factor using one of the following methods based on the number of terms:

   i. 4 terms: Factor by grouping.

   ii. 3 terms: Factor by using the reverse FOIL strategy. If the leading coefficient is 1, you can do this by identifying two factors that multiply together to give you the last term and add together to give you the middle term.

   iii. 2 terms: Use the difference of squares or the sum or difference of cubes. Remember that the sum of squares NEVER factors.

c. If none of these strategies works, you'll need to use the quadratic formula or identify the expression as not factorable (prime).

Remember that you can check any problem by multiplying the answer to ensure you get the original problem with which you started. Now that we have our general strategy, let's factor $6x^2 - 28x - 48$.

$2(3x^2 - 14x - 24)$ ←Factor out a greatest common factor (GCF) of 2.

Since $3x^2 - 14x - 24$ has three terms, we are going to use the reverse FOIL strategy. You can use a trial and error approach to factor this expression.

Possible factors of $3x^2 - 14x - 24$	Resulting product
$(3x+12)(x-2)$	$3x^2 + 6x - 24$
$(3x+8)(x-3)$	$3x^2 - x - 24$
$(3x-4)(x+6)$	$3x^2 + 14x - 24$

Notice that the last combination gives us what we want, just with the wrong sign on the middle term. So, if we switch the signs, we'll have what we want. Therefore, $3x^2 - 14x - 24$ factors into $(3x+4)(x-6)$, and the complete factorization of $6x^2 - 28x - 48$ is $2(3x+4)(x-6)$.

None of these terms will factor further, so this is your final answer.

2. Solve: $4x^2 - 27 = 9$.

To begin the problem, set one side equal to 0.

$4x^2 - 27 = 9$

$$\begin{array}{r} 4x^2 - 27 = 9 \\ -9 \quad -9 \\ \hline 4x^2 - 36 = 0 \end{array}$$ ← Subtract 9 from both sides. Simplify.

$4(x^2 - 9) = 0$ ←All terms on the left have a greatest common factor of 4, so factor out a 4.

$4(x-3)(x+3) = 0$ ←Factor $x^2 - 9$, which is a difference of perfect squares. The difference of squares always factors as the sum and difference of the two square roots.

$$\begin{array}{rr} x - 3 = 0 & x + 3 = 0 \\ +3 \quad +3 & -3 \quad -3 \\ \hline x = 3 & x = -3 \end{array}$$ ← Find the solutions by setting each factor equal to zero and solving for *x*.

$x = 3$, *or* $x = -3$.

TRY IT!

1. Factor: $x^2+7x+10$

2. Factor: x^2+5x-4

3. Solve: $x^2-11x-12=0$

Answers

1. Two numbers that multiply to 10 and add up to 7 are 5 and 2. So, the factors of the polynomial are $(x+2)(x+5)$.
2. This is unfactorable, since none of the factors of -4 (-1 and 4, -2 and 2, and -4 and 1) add up to 5.
3. Our strategy will be to factor the trinomial and then use the Zero Product Property. We can use -12 and 1 to factor the trinomial:

$x^2-11x-12=0$

$(x-12)(x+1)=0$

$x-12=0$ or $x+1=0$

$x=12$ or $x=-1$

The **roots**, solutions, or x-intercepts of the polynomial are 12 and -1.

3. The following are additional problems to review concerning reducing rational expressions and scientific notation. These are skills that will prove most useful in the Algebra II chapter that follows.

Simplify $\frac{(3x^2)^3}{(6x^3)^5}$

The following are the rules for simplifying exponents:

$x^a \cdot x^b = x^{a+b}$

$$\frac{x^a}{x^b} = x^{a-b}$$

$$(x^a)^b = x^{ab}$$

$$x^{-a} = \frac{1}{x^a}$$

This problem uses several of these rules. It's easiest to begin by raising the powers to the powers outside of the parentheses.

$$\frac{(3x^2)^3}{(6x^3)^5} = \frac{27x^6}{7776x^{15}}$$

Both 27 and 7,776 can be divided by 27, so we'll do that and subtract the exponents of the *x* terms:

$$\frac{1}{288}x^{-9}$$

Writing an answer with a negative exponent isn't considered standard, so if we put it in the denominator of the fraction, we can use a positive exponent. Therefore, our final answer becomes:

$$\frac{1}{288x^9}$$

4. Multiply (2.4 x 10^4)(5.4 x 10^{16}) and write the answer in scientific notation.

To use scientific notation, multiply the numbers and then find the factor of 10 using the rules above for exponents.

2.4 x 5.4 = 12.96

10^4 x 10^{16} = 10^{20}

You're almost done. The only remaining step is to convert 12.96 to scientific notation, which is 1.296 x 10, then multiplying by 10^{20} = 1.296 x 10^{21}.

Terms You Should Know

Direct Variation - When the ratio of two variables is constant.

Factor - A number that evenly divides into another number.

Greatest Common Factor (GCF) - The largest factor of two or more numbers or terms.

Inverse Variation - Represented by the model $y = \frac{k}{x}$ where *k* is the constant of variation.

Isolating the Variable - Using inverse operations to undo addition, subtraction, multiplication, and division in order to get the variable alone.

Linear Equations – An equation whose graph is a line.

Roots - The *x*-intercepts are also known as the roots of the equation. They are the points where the graph intersects the *x*-axis.

Slope – The rate of change of a line. For any two points, it is the change in the *y* values divided by the change in the *x* values, and is referred to as the change in the rise over the change in the run of a line.

Solving an Equation - Finding the value of the variable.

Variables - A letter used to represent an unknown value.

X-Intercept - The point at which the graph of an equation crosses the x-axis.

Y-Intercept - The point at which the graph of an equation crosses the y-axis.

Algebra II

Algebra, as stated earlier, uses arithmetic, logical applications of arithmetic, and several symbols to allow you to solve problems. Algebra problems involve a missing value or a value that can change when conditions change. We call the value we are searching for in algebraic problems a **variable.** You translate problems in algebra, and specifically in this section, Algebra II, into expressions and equations. An **expression** is a mathematical statement that does not contain an equals sign. An **equation** is a set of two expressions combined with an equals sign.

Algebra II courses focus on the relationships between two or more linear equations, usually called **systems of equations**. You can solve these problems in a variety of ways–with algebra and/or graphs. Algebra II classes also include absolute value and quadratic, exponential, and logarithmic functions. Finally, courses in Algebra II also spend a tremendous amount of time analyzing conic sections–the parabola, circle, ellipse, and hyperbola. Algebra II courses give students the opportunity to advance their understanding of algebra and hone their skills in algebra as a preparation for pre-calculus and calculus courses.

In an Algebra II course, you should begin to focus on looking at a problem, analyzing it before you begin, and trying to take the best approach to solving the problem, which is the approach that leads to the solution in the most efficient manner possible. Sometimes this isn't possible and you may need to try one or two different methods for solving a problem before you end up at the appropriate solution. Thus, Algebra II is a process that may require trial and error. Don't be afraid to make an error and learn from it!

Example 1: Systems of Linear Equations

Find the solution for the system of linear equations: $\begin{matrix} x - y = -3 \\ 2x - y = -7 \end{matrix}$

There are four different methods that can be used to solve a system of linear equations: **substitution**, **elimination**, **graphing**, and **matrices**. We'll solve this problem using each of the four methods to show that you get the same answer regardless of which method you choose; then we'll talk about the pros and cons of each method. Before we do that, it's important to understand what it is you are finding when you are solving a system of linear equations. The solution to a system of linear equations is the set of points that the linear equations have in common. The point in common is the *x* and *y* value that works in both equations (or the intersection of the two lines). You can have one of three things happen when you solve a system of linear equations: a single point can result, meaning the two lines intersect exactly one time (which is most common); no solution can occur, which happens when the two lines are parallel so they have no points in common; or you can get an infinite number of solutions, which means that the two equations are really the same line and the equations are written a bit differently.

Method 1: Substitution

To use substitution to solve a system of equations, you must get one of the equations into a form that can be substituted into the other equation for one of the variables. In this situation, the first equation ($x-y=-3$) is the best choice to rearrange for substitution, because it can quickly be rewritten into an *x* = or *y* = format. Let's get this equation into *x* = format by solving for *x*.

$$x-y=-3$$

$$\begin{array}{r} x - y = -3 \\ +y \quad +y \\ \hline x = -3 + y \end{array}$$

← To solve for *x* we need to add *y* to both sides, and simplify.

Now that we have the equation solved for one of the variables, we can substitute this into the other equation so that it only has one variable in it. So, we're going to write $-3+y$ in the first equation everywhere we see an *x*.

$$2x-y=-7$$

$2(-3+y)-y=-7$ ←Substitute -3 + *y* for *x*.

$-6+2y-y=-7$ ←Distribute the 2 through the parentheses.

$-6+y=-7$ ← Simplify.

$$\begin{array}{l} -6+y=-7 \\ \underline{+6 \qquad +6} \\ y=-1 \end{array}$$

←Add 6 to both sides to solve for *y*. Simplify.

Now that you have found *y*, you must substitute back into one of the equations and use what you know *y* to be in order to solve for *x*. We'll substitute *y* = -1 into the first equation that we solved for *x*.

$x=-3+y$ ← This was the equation we found when we solved $x-y=-3$ for *x*.

$x=-3+(-1)$ ← Substitute -1 in for *y*.

$x=-4$ ← Simplify.

Now that we know both *x* and *y*, we can write the answer as an ordered pair (-4, -1). Now we'll use the elimination method to solve the same problem and see that we get the same answer.

Method 2: Elimination

When you use elimination to solve a system of equations, the goal is to find a number by which you can multiply one equation (or numbers by which you can multiply each equation) so that when you add the two equations together, one of the variables will be eliminated from the equation. Let's take a look at the system we are solving again to see how this works.

$$x-y=-3$$
$$2x-y=-7$$

In this example, notice that the *y* terms in both equations are the same. If we multiply one of the equations (it doesn't matter which one) by -1, we'll be able to add the two equations together in order to "eliminate" the *y* terms so we can then solve for *x*.

$$\begin{array}{lll} x - y = -3 \longrightarrow & -1\,(x - y = -3) \xrightarrow{\text{multiply by} -1} & -1x - y = 3 \\ 2x - y = -7 \longrightarrow & 2x - y = -7 \longrightarrow & +\,2x - 7 = -7 \\ & & \overline{1x \qquad = -4} \end{array}$$

Now that you've found *x*, substitute this figure back into one of the equations and solve for *y*. We'll substitute it into the second equation and solve for *y*.

$2(-4) - y = -7$ ←Substitute -4 in to the equation for *x*.

$-8 - y = -7$ ← Simplify.

$$\begin{array}{rr} -8 - y = & -7 \\ +8 \qquad & +8 \\ \hline -y \qquad = & 1 \end{array}$$

← Add 8 to both sides. Simplify.

$\frac{-y}{-1} = \frac{1}{-1}$ ←Divide both sides by -1 to solve for *y*.

$y = -1$ ←Simplify.

THINK: You also could have multiplied the first equation by -2 to eliminate the *x* term as well. Would that have changed the answer? No. The answer should be the same regardless of the method or elimination steps.

Again, now that we have solved for x and y, we can write the point as an ordered pair:

(-4, -1).

Elimination is a nice method to use in a problem like this one where a quick multiplication of one or both of the equations will allow you to quickly eliminate one of the variables and solve for the other variable. Elimination is also nice to use when you have fractions or decimals in a problem, as you can multiply by a common denominator or power of 10 to clear the fractions or decimals and then multiply by the number needed to eliminate one of the variables.

Method 3: Graphing

The third method of solving a system of linear equations is to use graphing. Graphing, prior to the use of graphing calculators, was considered a "primitive" method of solving a system of equations because you can graph the lines quickly by putting them in slope-intercept form or by finding a few points on the lines and sketch a graph. However, the point at which they intersect on the graph might not be precise when you sketch by hand.

Now that graphing calculators are more prevalent, graphing has become a relatively valid method of solving systems of linear equations. Most graphing calculators or graphing software include a way to find the intersection of two lines by looking at the graph of the lines.

Let's take a look at how to locate the solution of the system of equations when you graph the two lines by hand. Recall that to graph a line, it is easiest to do this when the line is in $y = mx + b$ form, where m is the slope of the line and b is the y-intercept of the line. (See page 52 for a review). We'll begin by putting the equations into this form by solving each for y.

$$x - y = -3 \quad \rightarrow \quad -y = -x - 3 \quad \rightarrow \quad y = x + 3$$
$$2x - y = -7 \quad \rightarrow \quad -y = -2x - 7 \quad \rightarrow \quad y = 2x + 7$$

These two equations are relatively easy to graph. Recall that to graph $y = x + 3$, you put a point at the y-intercept, which is (0, 3).

Then you use the slope (which in this case is $\frac{1}{1}$) to go up 1 and then right 1 from that point and locate another point. This gives you the following graph:

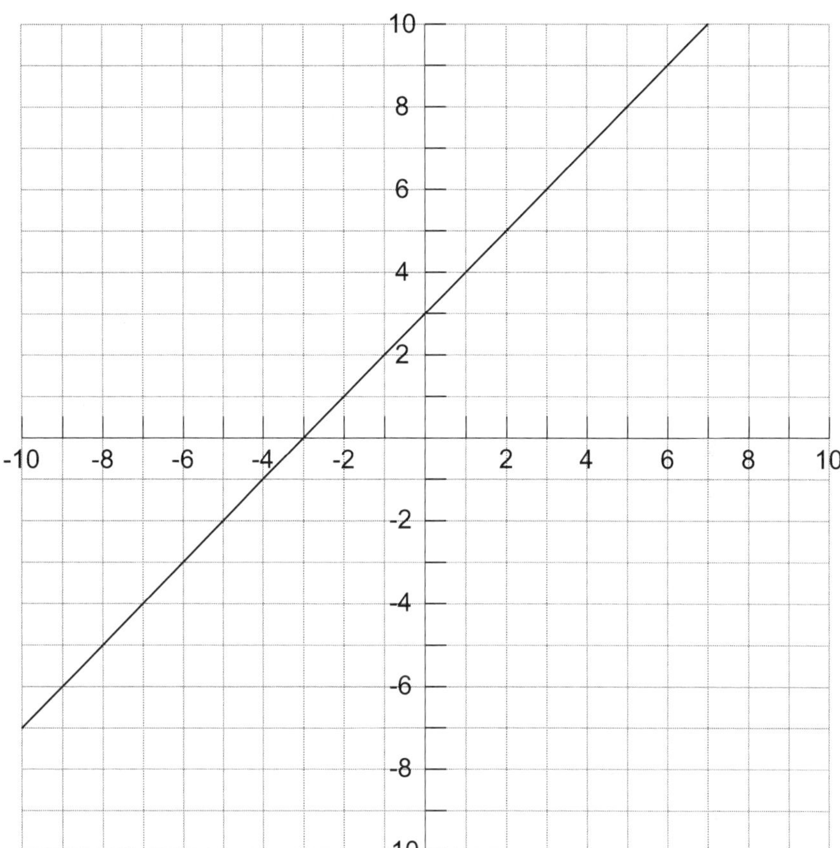

Now we want to graph the other line with this first line. We graph that equation by putting a point at (0, 7) and then using the slope of $\frac{2}{1}$ to plot an additional point that is 2 spaces up and 1 space to the right of this point. When we graph both lines together, our graph will look like the figure below:

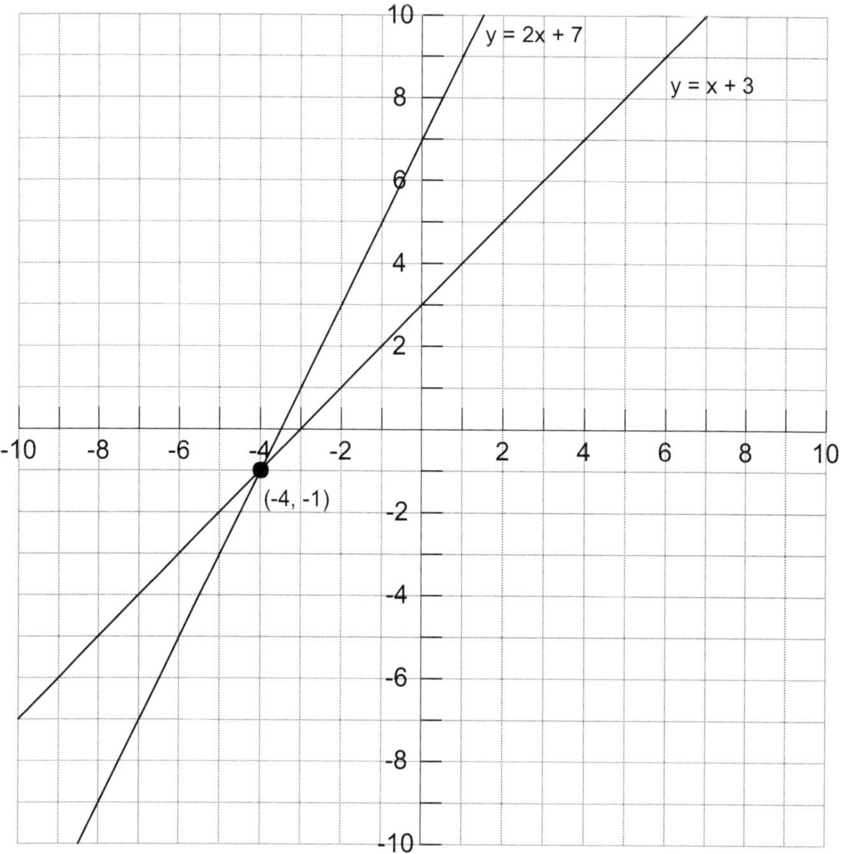

You can see that the solution appears to be what we expect, which is (-4, -1). On a graphing calculator, you could confirm this by using the "intersect" calculation on the calculator. If you were graphing by hand, you should check this point in both problems to see that the point is really on both lines.

THINK: If two lines are parallel, will they have a point of intersection? No, they will not and thus there will be no solution to the system. If two lines are the same or coinciding, will they have a point of intersection? Yes, they will have infinite points of intersection or infinite solutions.

Method 4: Matrices

The fourth method for solving a system of linear equations is to use matrices. A matrix is an arrangement of numbers into rows and columns. Matrices have become a more popular method for solving a system of equations with the growing use of graphing calculators and mathematical computer software, because most graphing calculators and computer software programs provide an easy way to perform operations with matrices and to find values of matrices without having to do long, difficult calculations. If you are not using a graphing calculator or computer software, matrices are often a poor choice, as calculations on matrices can be very long and complex—far more difficult for a system of linear equations than substitution or elimination. However, when you move beyond systems of linear equations to equations with three or more variables, matrices are often far more effective than substitution and elimination.

There are two methods for solving a system of equations using matrices; one is using row elimination to simplify the system, and the other is to use a technique called **Cramer's Rule**. Cramer's Rule is the method employed when you're using a graphing calculator to solve a system of equations—this also can be done pretty easily without a calculator. We'll demonstrate the process of row elimination here and explain the basic process of Cramer's Rule using a calculator or computer software. You will need to consult the manual or Help files for your particular calculator or computer software to find out how to perform these specific calculations.

Matrices Using Row Elimination by Hand

To use row elimination, you must first put your system of equations into a matrix. Let's look again at our system of equations and how it would be written as a matrix.

$$\begin{array}{l} x - y = -3 \\ 2x - y = -7 \end{array} \quad \Rightarrow \quad \left[\begin{array}{cc:c} 1 & -1 & -3 \\ 2 & -1 & -7 \end{array}\right]$$

As you can see, the coefficients (or numbers multiplied by the variables) of the two equations are written in horizontal rows. There is also a dotted or dashed line that is written before the column that contains the values on the other side of the equals sign. Now we want to use row-equivalent operations to re-write the matrix in a new way. There are three row-equivalent operations. They are:

a) Interchanging any two rows

b) Multiplying each element of a row by the same non-zero number

c) Multiplying each element of a row by the same non-zero number and adding the result to another row.

Using these three operations, the goal is to get the bottom left element equal to 0. When we do this, we can then write a simple equation and solve for *y*. Then we'll substitute that value of *y* into one of our original equations and solve for *x*.

So, using the row-equivalent operations, we need to turn the 2 that is in the bottom left corner into a 0. There are many ways to do this, but the easiest appears to be multiplying each element in the top row by -2 and then adding this to the second row. We'll re-write this new row in the second row's place. So, let's rewrite the top row, multiplied by -2, and then fill in the second row.

$$\left[\begin{array}{cc:c} -2 & 2 & 6 \\ & & \end{array}\right]$$

To find the values for the second row, you'll add the new top row (-2, 2, and 6) to the elements of the original bottom row (2, -1, and -7). When you add -2 and 2 you get 0, when you add 2 and -1 you get 1, and when you add 6 and -7 you get -1, so your new bottom row is (0, 1, and -1). Let's take a look at that.

$$\left[\begin{array}{cc:c} -2 & 2 & 6 \\ 0 & 1 & -1 \end{array}\right]$$

This tells us that 1*y* = -1 or *y* = -1. Now, we can use this to substitute into one of the original equations and solve for *x*. We'll use the equation $x - y = -3$.

$x - y = -3$

$x - -1 = -3$	←Substitute.
$x + 1 = -3$	←Simplify.
$x + 1 - 1 = -3 - 1$	←Subtract 1 from both sides.
$x = -4$	←Simplify. Write your answer as an ordered pair.

Matrices Using a Graphing Calculator

Cramer's Rule involves finding the determinant of a matrix.

The **determinant** $\begin{vmatrix} a_1 & b_1 \\ a_2 & b_2 \end{vmatrix}$ is defined to mean $a_1b_2 - a_2b_1$.

A determinant can be found for any size matrix, so Cramer's Rule will work for any system of equations. For example, if you have three equations with three variables,

the determinant is: $$\begin{vmatrix} a_1 & b_1 & c_1 \\ a_2 & b_2 & c_2 \\ a_3 & b_3 & c_3 \end{vmatrix} = a_1\begin{vmatrix} b_2 & c_2 \\ b_3 & c_3 \end{vmatrix} - a_2\begin{vmatrix} b_1 & c_1 \\ b_3 & c_3 \end{vmatrix} + a_3\begin{vmatrix} b_1 & c_1 \\ b_2 & c_2 \end{vmatrix}.$$

To use the determinant to solve a system of linear equations, you have to find three different determinants.

In a general system of equations $\begin{matrix} a_1x + b_1y = c_1 \\ a_2x + b_2y = c_2 \end{matrix}$ we find three determinants, which we call D, D_x, and D_y.

$$D = \begin{vmatrix} a_1 & b_1 \\ a_2 & b_2 \end{vmatrix}$$

$$D_x = \begin{vmatrix} c_1 & b_1 \\ c_2 & b_2 \end{vmatrix}$$

$$D_y = \begin{vmatrix} a_1 & c_1 \\ a_2 & c_2 \end{vmatrix}$$

Cramer's Rule says $x = \frac{D_x}{D}$ and $y = \frac{D_y}{D}$.

The easiest way to solve a system of equations on your graphing calculator or with computer software is to enter the three matrices (for D, D_x, and D_y) into the calculator or software, use the calculator or software to find each determinant, and then do the simple division problems shown above to get *x* and *y*. **Note**: With a system of linear equations in two variables, it's fairly easy to calculate by hand, too.

You also might try writing a program on the graphing calculator or downloading a program off the Internet that will perform the calculations in Cramer's Rule. Many are available free of charge from a variety of websites. It is important to note that most Algebra II classes will require mastery of substitution, graphing, and elimination.

Exercise 1

Solve the system of equations by substitution.

$$4x - 3y = 15$$
$$x + 3y = 0$$

To substitute, remember you want to solve one equation for one of the variables and then substitute this into the other equation. It will be easiest to solve the second equation for *x*.

$$x + 3y = 0$$

$x + 3y - 3y = 0 - 3y$ ←Subtract 3*y* from both sides.

$x = -3y$ ←Simplify.

Now we can substitute $-3y$ in for *x* in the first equation.

$4(-3y) - 3y = 15$ ←Substitute.

$-12y - 3y = 15$ ←Simplify.

$-15y = 15$ ←Combine like terms.

$\frac{-15y}{-15} = \frac{15}{-15}$ ←Divide by -15 to get *y* by itself.

$y = -1$ ←Simplify.

Now that you know $y = -1$ you can substitute that back into one of your equations to solve for *x*. We'll substitute it back into *x* + 3*y* = 0.

$x + 3(-1) = 0$ ←Substitute.

$x - 3 = 0$ ←Simplify.

$x - 3 + 3 = 0 + 3$ ←Add 3 to both sides to solve for *x*.

$$x = 3$$

So the solution of this system of equations is (3, -1).

Exercise 2

Solve the system of equations by elimination.

$$3x - 4y = 6$$
$$5x + 9y = 10$$

Remember that the goal in elimination is to multiply one equation by a number or each equation by numbers so that when you add the two equations together, one of the variables is eliminated. In this problem, it appears we'll have to multiply both equations by a number. Since the *y* term in the first equation is negative and it's positive in the second, let's choose numbers so that we can eliminate the *y* term. If we multiply the top equation by 9 and the bottom equation by 4, when we add them together the *y* terms should be eliminated.

$$\begin{array}{lcr} 9(3x - 4y = 6) & \rightarrow & 27x - 36y = 54 \\ 4(5x + 9y = 10) & \rightarrow & \underline{20x + 36y = 40} \\ & & 47x = 94 \end{array}$$ ←Multiply and add the two equations.

$$\frac{47}{47}x = \frac{94}{47}$$ ←Divide both sides by 47.

$$x = 2$$ ←Simplify.

Now that we know that $x = 2$, we can use this to find *y* by substituting it into one of our equations. Let's use the top equation.

$$3(2) - 4y = 6$$ ←Substitution.

$6 - 4y = 6$ ←Simplification.

$6 - 6 - 4y = 6 - 6$ ←Subtract 6 from both sides.

$$-4y = 0$$

$$\frac{-4y}{-4} = \frac{0}{4}$$ ←Divide both sides by -4.

$y = 0$

So, the solution of this system is (2, 0).

Exercise 3

Solve the system of equations.

$$3x - 2y = 6$$
$$6x - 4y = 8$$

We can use any method we like to solve this system, so let's use elimination. If we multiply the top equation by -2, we will eliminate the *x* terms.

$$\begin{aligned} -2(3x - 2y = 6) &\Rightarrow -6x + 4y = -12 \\ 6x - 4y = 8 &\Rightarrow \underline{\;\;6x - 4y = \;\;8\;\;} \\ & \qquad\quad\; 0 \;=\; -4 \end{aligned}$$

Something odd happened here. Instead of just one of the variables being eliminated they were both eliminated. Notice that when we added the two equations together, we got a false statement (0 does not equal -4). Since we got a false statement, it means there is no solution. If you looked at a graph of these two lines, the lines would be parallel to one another.

Exercise 4

Solve the system of equations.

$$y = 3x - 2$$
$$6x - 2y = 4$$

Again, we're not told how to solve this problem. Since the top equation is already solved for *y*, let's use substitution. So, we'll substitute the top equation into the bottom equation in place of *y*.

$6x - 2(3x - 2) = 4$ ←Substitute.

$6x - 6x + 4 = 4$ ←Distribute the -2 through the parentheses.

$4 = 4$ ←Simplify.

Notice that again both variables cancelled out. However, in this example, we got a true statement as our result (4 equals 4). This means any solution that is a solution of one of the equations will be a solution to the other equation. If you looked at a graph of these two equations, they would appear to be the same line.

TRY IT!

Solve each system of linear equations through substitution, elimination, and graphing.

1. $\begin{array}{l} y = 3x + 1 \\ y = x + 5 \end{array}$

2. $\begin{array}{l} x + 2y = 5 \\ -x + y = 13 \end{array}$

Answers

1. Substitution.

Here both equations have isolated the *y*-value, so we will begin by substituting one equation into the other for the value of *y*.

$y = 3x + 1$
$y = x + 5$

$3x + 1 = x + 5$	← Substitute 3x+1 for *y* in the second equation.
$2x + 1 = 5$	←Combine like terms by subtracting *x* from both sides.
$2x = 4$	← Combine the constant terms.
$x = 2$	←Simplify.

Now that you've found *x*, substitute this figure back into one of the equations and solve for *y*. We'll substitute it into the first equation and solve for *y*.

$y = 3x + 1$	←The original equation.
$y = 3(2) + 1$	← Substitute 2 in for *x*.
$y = 7$	← Simplify.
$(2,7)$	←Solution.

2. Elimination

$y = 3x + 1$
$y = x + 5$

$$
\begin{array}{llll}
y = 3x + 1 \longrightarrow & -3x + y = 1 & \xrightarrow{\text{multiply by } -1} & 3x - y = -1 \\
y = x + 5 = \longrightarrow & -x + y = 5 + & \longrightarrow & -x + y = 5 \\
 & & & 2x = 4 \quad \text{add the two equations} \\
 & & & x = 2
\end{array}
$$

Next we substitute the value of *x* into an equation to solve for *y*.

$y = 3(2) + 1$
$y = 7$

So, the coordinate of intersection and the solution for this system is $(2,7)$.

3. Graphing

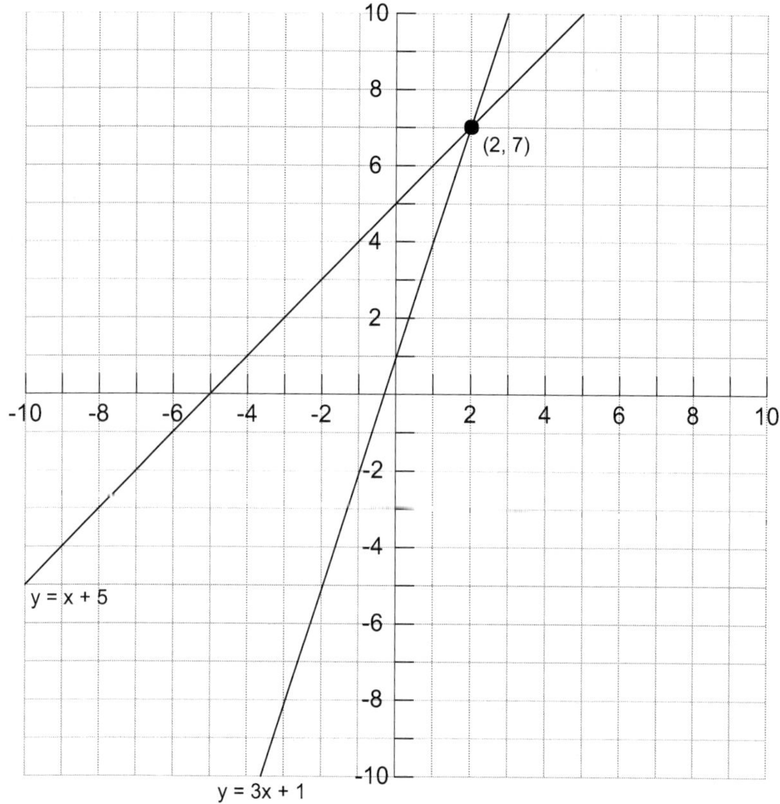

2. $\begin{aligned} x+2y&=5 \\ -x+y&=13 \end{aligned}$

1. Substitution

Here we have to choose an equation and isolate the *x* or the *y*. In this system of equations, the *x* is without a coefficient in the first equation and the *y* is in the second equation. These are good candidates for isolating and solving for this substitution problem.

$$x+2y=5 \rightarrow x=-2y+5$$
$$-x+y=13$$

Now that we have solved for *x* in the first equation, we can use it for the value of *x* in the second equation as a substitution.

$$-(-2y+5)+y=13$$
$$2y-5+y=13$$
$$3y-5=13$$
$$3y=18$$
$$y=6$$

← Substitute -2y+5 for *x* in the second equation.

Now that you've found *y*, substitute this figure back into one of the equations and solve for *x*. We'll substitute it into the first equation and solve for *x*.

$x+2y=5$ ←The original equation.

$x+2(6)=5$ ← Substitute 6 in for *y*.

$x=-7$ ← Simplify.

$(-7,6)$ ←Solution.

1. Elimination

$$x+2y=5$$
$$-x+y=13$$

Since the coefficients of *x* are 1 and -1, we can quickly eliminate *x* from the system by adding the equations together.

$$\begin{array}{rr} & x+2y=5 \\ + & -x+y=13 \\ \hline & 3y=18 \\ & y=6 \end{array}$$

We can then substitute $y = 6$ into either equation to solve for x:

$$x+2(6)=5$$
$$x+12=5$$
$$x=-7$$

Finally, we can check our solution by substituting x and y into the equation we didn't use for our previous substitution:

$$-x+y=13$$
$$-(-7)+6=13$$
$$7+6=13$$
$$13=13$$

Thus, our solution is $(-7,6)$.

2. Graphing

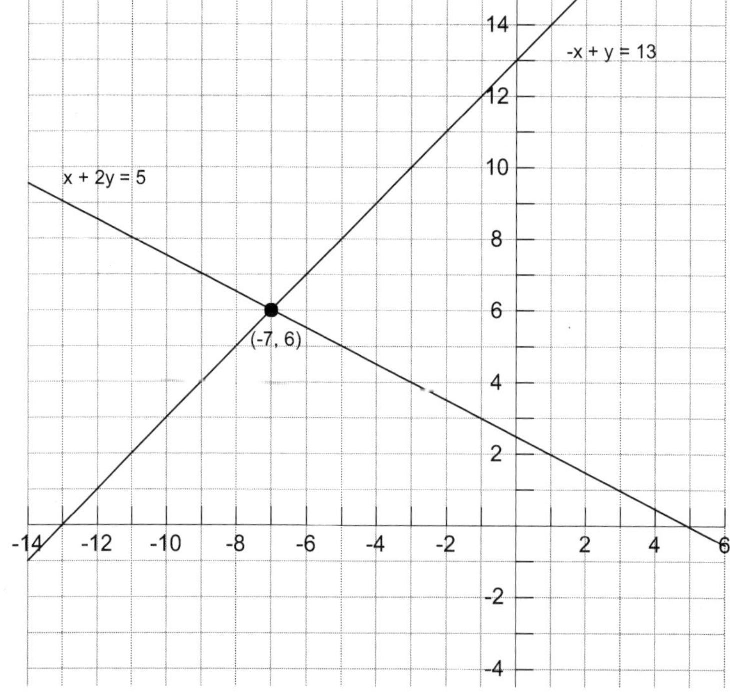

Example 2: Identifying a conic section from its equation

Determine, without graphing, what type of conic section is represented by this equation: $\frac{(x-3)^2}{4} - \frac{(y+1)^2}{1} = 1$.

Before we can solve this problem, we need to know more about conic sections. A **conic section**, which we often simplify by just saying a "conic," is formed when a cone or, in the case of the hyperbola, two cones stacked on their tips, are intersected by a plane. There are four different conic sections: the parabola, the circle, the ellipse, and the hyperbola. Let's take a look at a sample graph and the equations of each of these.

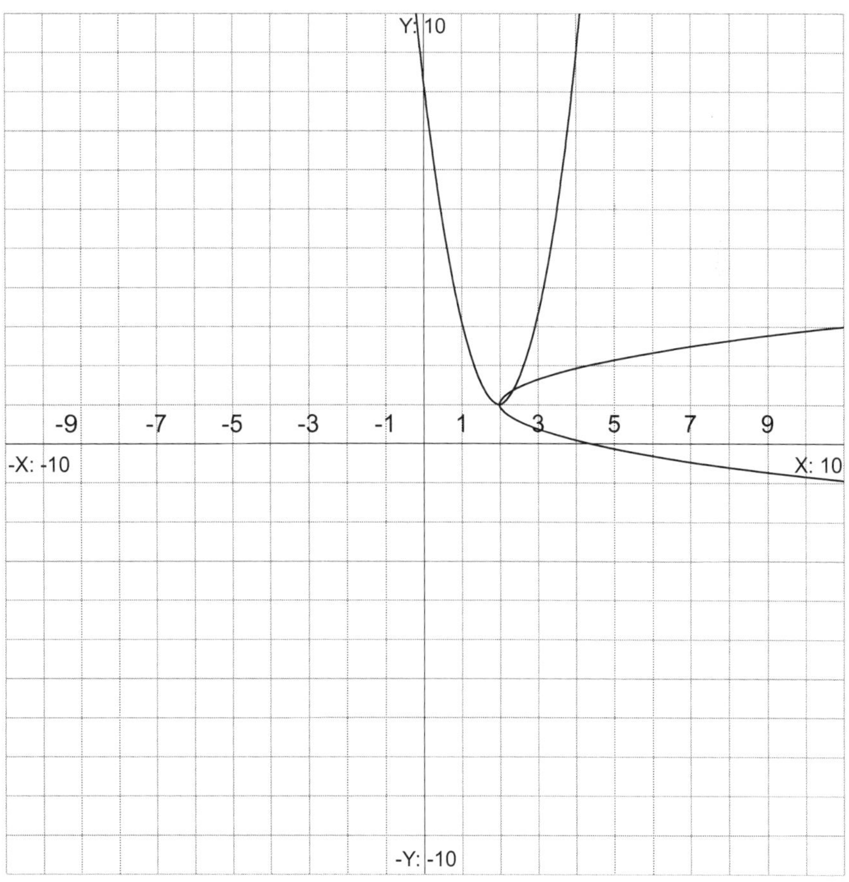

The Parabola

The parabola can open upward or downward (like the one shown in red in the graph to the left) or it can open to the left or right (like the one shown in green on the right).

The general form of the equation of a parabola that opens up and down is

$y = a(x-h)^2 + k$, where (h, k) is the vertex of the parabola. The vertex of a parabola is the point where it begins to turn back around. In a parabola that opens up and down, it is the bottom or topmost point, respectively.

The general form of parabola that opens to the left or right is $y = a(x-h)^2 + k$, where the vertex is (h, k).

Note: If you need to, review the discussion on factoring on pages 52 through 54. It will help you remember how to manipulate an equation of a parabola into standard form.

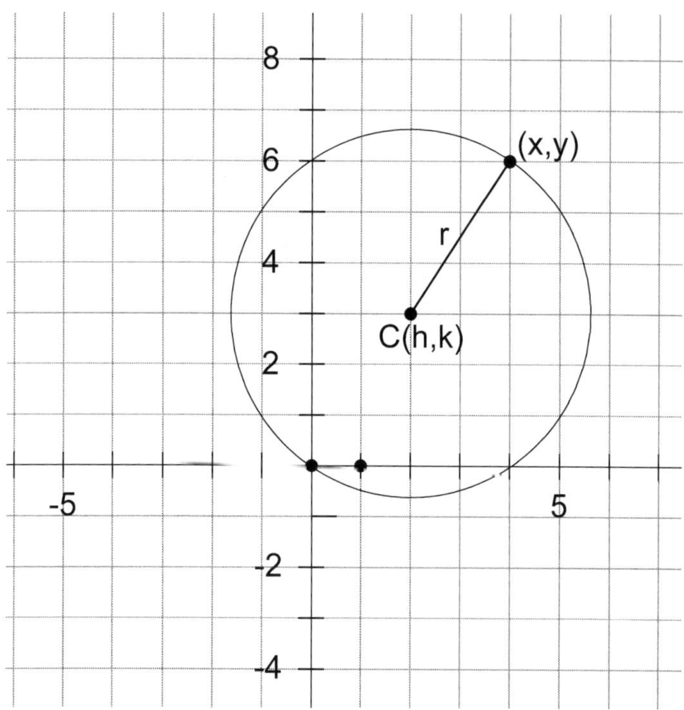

Circle

A circle has the general equation: $(x-h)^2+(y-k)^2=r^2$, where (*h*, *k*) is the center of the circle and *r* is the radius of the circle.

THINK: Notice the two variables, *x* and *y*, are squared, unlike the equation of a parabola where either the *x* or the *y* is squared (not both). This is a good hint when trying to identify conic sections from their equations.

Ellipse

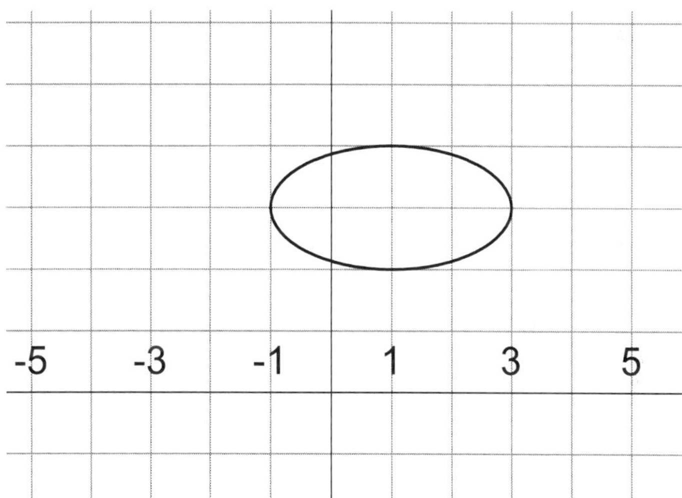

An ellipse has the general equation: $\frac{(y-k)^2}{a^2}-\frac{(x-h)^2}{b^2}=1$,

where (*h*, *k*) is the center of the ellipse. An ellipse has two special points called the foci. The foci are located at $(h\pm c,k)$, where $c^2=a^2-b^2$. An ellipse also can be taller than it is long. In that case,

the ellipse has the general equation $\frac{(y-k)^2}{a^2}+\frac{(x-h)^2}{b^2}=1$ with foci at $(h,k\pm c)$.

THINK: Notice the format of the equation for an ellipse, where the *x* and the *y* are squared and the numerator is a perfect square.

Hyperbola

The hyperbola that opens to the left and right (like the one shown in the figure below) has the general equation $\frac{(x-h)^2}{a^2}-\frac{(y-k)^2}{b^2}=1$. If the hyperbola opens up and down, then it has the general equation $\frac{(y-k)^2}{a^2}-\frac{(x-h)^2}{b^2}=1$

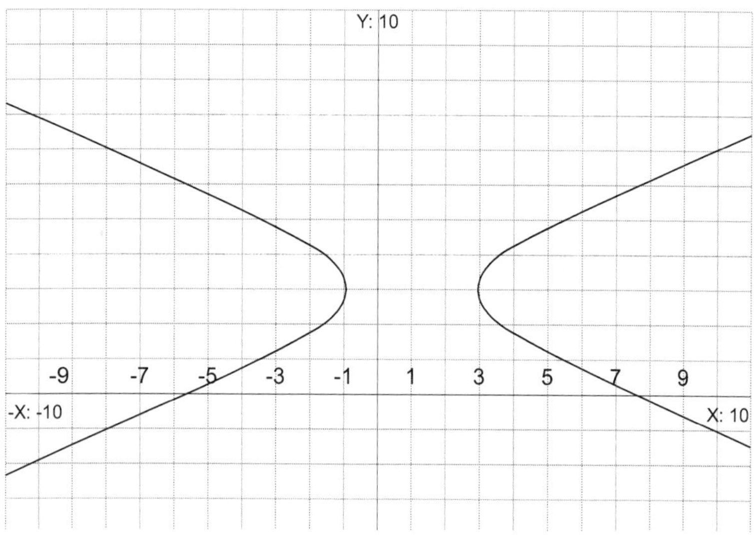

Now that we know a little bit about each of the types of conic sections, we need to see with which of them our equation matches up the best. Remember, our equation was $\frac{(x-3)^2}{4}-\frac{(y+1)^2}{1}=1$. Let's look at each of the types of conics and see if it fits that type of equation.

THINK: An ellipse has two rational expressions being added, and a hyperbola has two rational expressions being subtracted.

Parabola → This is not a parabola, because in a parabola either the *x* is squared or the *y* is squared, but never both.

Circle → This is not a circle, because the general form of a circle does not have anything that is dividing into each of the *x* and *y* terms. Also, the general form of the circle uses a plus sign between the two *x* and *y* terms, and this equation contains a minus sign.

Ellipse → This looks much like the general form of an ellipse, except the minus sign is used between the two terms, so it must not be an ellipse.

Hyperbola → This equation is not any of the other three conics, and it does match up with the general form of a hyperbola, so it must be a hyperbola.

Exercise 1

Find the equation of a parabola with vertex at (3, 4) and a point on the parabola at (5, 16).

Remember that the general formula for a parabola is $y = a(x-h)^2 + k$, where (*h*, *k*) is the vertex of the parabola. We are given the vertex and a point (*x*, *y*) so we will need to substitute these values in and solve for *a*.

$y = a(x-h)^2 + k$

$16 = a(5-3)^2 + 4$ ←Substitute the vertex (3,4) in for (*h*, *k*) and the point (5,16) in for (*x*, *y*).

$16 = 4a + 4$ ←Simplify.

$16 - 4 = 4a - 4$ ←Subtract 4 from both sides.

$12 = 4a$

$\frac{12}{4} = \frac{4a}{4}$ ←Divide both sides by 4.

$3 = a$ ←Simplify.

Now re-write the equation. When writing the equation of the parabola, fill in the values for *a*, *h*, and *k*. This gives us $y = 3(x-3)^2 + 4$.

Exercise 2

Find the equation of the circle with center at (2, -1) and with a diameter of 16.

Recall that the general form for the equation of a circle is $(x-h)^2+(y-k)^2=r^2$, where (*h*, *k*) is the center of the circle and *r* is the radius of the circle. We are given the center, but instead of being given the radius we are given the diameter. Remember that the radius is equal to half of the diameter, so we know that *r* is 8. Now, we just substitute in the values for *h*, *k*, and *r*. This gives us $(x-2)^2+(y+1)^2=8^2$.

We'll simplify that so we get a final answer of $(x-2)^2+(y+1)^2=64$.

TRY IT!

Decide if the equation is a parabola, hyperbola, ellipse, or a circle.

1. $-3x^2+y+2=0$
2. $y=-4x+2$
3. $\frac{x^2}{4}+\frac{y^2}{1}=1$
4. $\frac{(x-9)^2}{16}-\frac{(y+4)^2}{36}=1$
5. $(x-25)^2+(y-2)^2=100$

Answers

1. Since only the *x* is squared, this is a parabola.
2. This is not a conic section; it is a line. Notice the variables are raised to the first power.
3. Since the two rational expressions are being added, this is an ellipse.

4. Since the two rational expressions are being subtracted, this is a hyperbola.

5. This is the equation of a circle with center (25, 2) and radius 10.

Example 3: Logarithmic Functions

Find $\log_2 32$.

Let's first look at how we would read this problem. This would be read "log, base 2, of 32." That means if 2 is a base, to what power do we have to raise 2 to get 32? We could write this question as an equation. It would look like this:

$2^x = 32$.

If you know your powers of 2, this is relatively easy to solve. If you aren't sure though, just use trial and error until you get the right answer. Let's take a look at the powers of 2 until we find the right answer:

$2^1 = 2$

$2^2 = 4$

$2^3 = 8$

$2^4 = 16$

$2^5 = 32$

Therefore, $\log_2 32 = 5$.

The following definition and rules can be used when evaluating, simplifying, and working with logarithmic functions.

1. Converting between exponential and logarithmic equations:

   $y = \log_a x \longleftrightarrow a^y = x$

   Remember that the variable *a* in this form represents what we call the base. (To review the exponent rules, go to page 70.)

2. For any base *a*, $\log_a 1 = 0$. The logarithm, base *a*, of 1 is always

0. THINK: $a^0 = 1$

3. For any base *a*, $\log_a a = 1$ THINK: $a^1 = a$

4. The product rule: For any positive numbers *M* and *N*, $\log_a(M \cdot N) = \log_a M + \log_a N$

5. The power rule: For any positive number *M* and any real number k, $\log_a M^k = k \cdot \log_a M$

6. The quotient rule: For any positive numbers *M* and *N*, $\log_a M^k = k \cdot \log_a M$

7. The change of base rule: For any positive number *M* and any bases *a* and *b*, $\log_a M = \dfrac{\log_b M}{\log_b a}$

8. For any base *a*, $\log_a a^k = k$

9. For any base *a*, $a^{\log_a k} = k$

Exercise 1

Given that the $\log_a 3 = 0.201$ and the $\log_a 5 = 0.681$, find each of the following:

a) $\log_a 15$

b) $\log_a 25$

c) $\log_a \frac{5}{3}$

Recall from Example 3 above the following rules for logarithmic functions:

The product rule: For any positive numbers *M* and *N*, $\log_a(M \cdot N) = \log_a M + \log_a N$

The power rule: For any positive number *M* and any real number *k*, $\log_a M^k = k \cdot \log_a M$

The quotient rule: For any positive numbers *M* and *N*,

$$\log_a \frac{M}{N} = \log_a M - \log_a N$$

Since we know the values of $\log_a 3$ and $\log_a 5$, let's try to rewrite the given log problem in terms of $\log_a 3$ and/or $\log_a 5$. Let's begin with the first one:

a) $\log_a 15$

$= \log_a(3 \cdot 5)$ ← Rewrite 15 in terms of 3 and 5, the two logs we know.

$= \log_a 3 + \log_a 5$ ←Rewrite this using the product rule.

$= 0.201 + 0.681$ ←Substitute the two values we were given.

$= 0.882$

b) $\log_a 25$

$= \log_a 5^2$ ←25 is equal to 5^2.

$= 2\log_a 5$ ←Rewrite using the power rule.

$= 2(0.681)$ ←Substitute the value for $\log_a 5$.

$= 1.1362$

c) $\log_a \frac{5}{3}$

$= \log_a 5 - \log_a 3$ ←Rewrite using the quotient rule.

$= 0.681 - 0.201$ ←Substitute the values.

$= 0.48$ ←Simplify.

TRY IT!

1. Express the equation in exponential form: $\log_5 25 = 2$

2. Express the equation in a logarithmic form: $8^{-1} = \frac{1}{8}$

3. Evaluate the expression: $\log_4 64$

Answers

1. $\log_5 25 \ = 2$ in exponential form is $\log_5 25 \ = 2$

2. $\log_8 \frac{1}{8} = \ -1$

3. $4^x = 64$ in this case *x* must be 3.

Example 4: Operations on Rational Expressions

Multiply $\frac{x+4}{2+x} \cdot \frac{x^2 - x - 6}{x^2 - 16}$.

Recall from basic arithmetic that to multiply two fractions means to multiply numerator times numerator and denominator times denominator. For example:

$$\frac{x+4}{2+x} \cdot \frac{x^2 - x - 6}{x^2 - 16}$$

Also remember $\frac{6}{12}$ can be simplified to $\frac{1}{2}$ because 6 will divide evenly into both the numerator and denominator. You should also remember from your basic arithmetic study that you could have simplified the problem before working it by canceling the 3 in the numerator of the first fraction with the 3 in the denominator of the second fraction, which would have given us the problem:

$$\frac{3}{4} \cdot \frac{2}{3} = \frac{1 \cdot 2}{4 \cdot 1} = \frac{2}{4} = \frac{1}{2}$$

Likewise, the problem could have been simplified a bit more before solving it. You could have cancelled the 2 in the numerator of the second fraction with the 4 in the denominator of the first fraction which would have given us the problem:

$$\frac{3}{4}\cdot\frac{2}{3}=\frac{1\cdot 2}{4\cdot 1}=\frac{2}{4}=\frac{1}{2}$$

(See page 2 to review working with fractions.)

Note that simplifying first (like we did in the most recent example) is the quickest way to get the solution. The same is true when you are working with rational expressions like the ones in our problem. So, we'll begin this problem by trying to find out if there is anything we can cancel between the numerator and denominator of the two fractions. Let's take a look at our problem:

$$\frac{x+4}{2+x}\cdot\frac{x^2-x-6}{x^2-16}$$

The second fraction contains two polynomials that will factor. They factor like so:

$$\frac{x+4}{2+x}\cdot\frac{(x-3)(x+2)}{(x+4)(x-4)}$$

(For a review of factoring, see page 63.)

The *x* + 2 terms will both cancel as will the *x* + 4 terms. This leaves us with an answer of $\frac{x-3}{x-4}$.

Think: What are the restrictions on the equation? Consider the denominator of the expression, x - 4 and recall that a denominator ≠0. So, x - 4 ≠ 0 or x ≠ 4. So, *x* can equal all real numbers but 4.

Exercise 1

Solve $\frac{x+5}{x-2}=\frac{x-2}{x+4}$

Solving rational equations is one of the critical skills you will need in later math courses. Solving an equation like this one is relatively straightforward if you recall work you have done in the past with solving proportions. Before we start solving this problem, let's take a look at something similar but a bit easier from our past studies.

A proportion like $\frac{x}{2}=\frac{20}{5}$ is solved by cross-multiplying. When you cross-multiply you multiply the numerator of one fraction by the denominator of the other and set that equal to the product of the other denominator and numerator. To solve $\frac{x}{2}=\frac{20}{5}$ you would cross-multiply and get $5x = 40$. Dividing both sides by 5 would give you $x = 8$.

We solve this problem the same way. So, we'll cross-multiply and get:

$$(x+5)(x+4)=(x-2)(x-2)$$

Now, we need to multiply this out using FOIL. Remember that FOIL stands for First, Outer, Inner, Last, and reminds us how to multiply two binomials. We'll use FOIL to expand both sides of the equation above, starting with the left side: $(x+5)(x+4)$. First, we multiply the **f**irst terms in each binomial together (the *x* and the *x,* in this example). Second, we multiply the **o**uter-most terms together (the *x* and the *4*). Third, we multiply the **i**nner-most terms together (the *5* and the *x*). Finally, we multiply the **l**ast terms in each binomial (the *5* and the *4*). Repeat this process to expand the right hand side of the equation as well, and we get:

$$x^2+4x+5x+20=x^2-2x-2x+4$$

$x^2+9x+20=x^2-4x+4$ ←Combine like terms.

$x^2-x^2+9x+4x+20-4=0$ ←Move all terms to one side.

$13x+16=0$ ←Combine like terms.

$13x=-16$ ←Subtract 16 from both sides.

$\frac{13x}{13}=\frac{-16}{13}$ ←Divide by 13.

$x=-\frac{16}{13}$ ←Simplify.

TRY IT!

1. Simplify and give all restrictions: $\frac{x^2+5x-14}{8x^2}\cdot\frac{4x^2-12x}{3x+21}$

2. Simplify and give all restrictions: $\frac{15x^4}{9x}\cdot\frac{12x^5}{35x^3}$

3. Simplify and give all restrictions: $\frac{x^2-3x-10}{x^2-4x}\div\frac{x^2+4x+4}{x^2-16}$

Answers

1. Our strategy will be to factor the numerator and denominator of each fraction and then look for common factors.

$$\frac{x^2+5x-14}{8x^2}\cdot\frac{4x^2-12x}{3x+21}$$
$$=\frac{(x+7)(x-2)}{4x\cdot 2x}\cdot\frac{4x(x-3)}{3(x+7)}$$
$$=\frac{(x+7)\cdot 4x}{(x+7)\cdot 4x}\cdot\frac{(x-2)(x-3)}{2x\cdot 3}$$
$$=1\cdot\frac{(x-2)(x-3)}{6x}$$
$$=\frac{(x-2)(x-3)}{6x}$$

2. In this case, the restriction is $8x^2 \neq 0$, so, $x \neq 0$ and $3x+21\neq0$, so $x \neq -7$ (found by solving the inequality for x).

Our goal is to find the simplest possible expression for the product. To do this, we look for common factors in the numerator and denominator of the product. Remember that to multiply fractions, multiply their numerators and multiply their denominators.

$$\frac{15x^4}{9x} \cdot \frac{12x^5}{35x^3}$$
$$= \frac{15 \cdot 12 \cdot x^4 \cdot x^5}{9 \cdot 35 \cdot x \cdot x^3}$$
$$= \frac{3 \cdot 5 \cdot 3 \cdot 4 \cdot x^4 \cdot x^5}{3 \cdot 3 \cdot 5 \cdot 7 \cdot x^4}$$
$$= \left(\frac{3 \cdot 3 \cdot 5 \cdot x^4}{3 \cdot 3 \cdot 5 \cdot x^4}\right) \cdot \left(\frac{4 \cdot x^5}{7}\right)$$
$$= 1 \cdot \frac{4x^5}{7}$$
$$= \frac{4x^5}{7}$$

In this solution, the restrictions are $9x \neq 0$, so $x \neq 0$ and $35x^3 \neq 0$, so $x \neq 0$.

3. $\dfrac{x^2 - 3x - 10}{x^2 - 4x} \div \dfrac{x^2 + 4x + 4}{x^2 - 16}$

Before we solve the problem, we can find the restrictions on the solution by setting the expression in the denominator not equal to (≠) 0.

$$x^2 - 4x \neq 0 \qquad x^2 - 16 \neq 0$$
$$x(x-4) \neq 0 \qquad (x-4)(x+4) \neq 0$$
$$x \neq 0, x \neq 4 \quad \text{and} \quad x \neq 4, x \neq -4$$

$$\frac{x^2 - 3x - 10}{x^2 - 4x} \div \frac{x^2 + 4x + 4}{x^2 - 16}$$

$$= \frac{x^2 - 3x - 10}{x^2 - 4x} \cdot \frac{x^2 - 16}{x^2 + 4x + 4}$$
$$= \frac{(x-5)(x+2)(x+4)(x-4)}{x(x-4)(x+2)(x+2)}$$
$$= \frac{(x+2)(x-4)}{(x+2)(x-4)} \cdot \frac{(x-5)(x+4)}{x(x+2)}$$
$$= \frac{(x-5)(x+4)}{x(x+2)}$$

Using the solution we see x(x+2) ≠ 0, so x ≠ 0 or -2. Combining the restrictions above, x ≠ 0, -2, 4 and -4.

Example 5: Rational Roots

List all the possible rational roots of $4x^5 + 3x^4 - 2x^2 + 6x - 12 = 0$.

Before we can list the rational roots of this equation, we need to understand what a root is. A root of an equation is a number that will be a solution of the equation. For example, in the equation $x + 1 = 0$, -1 is a root of this equation because -1 can be substituted in for x and it gives a true answer (-1 + 1 = 0).

For a polynomial like the one we are working with, there is a theorem called the **Fundamental Theorem of Algebra,** which says an equation will have exactly as many roots as its highest degree. So, in this case, we will have exactly 5 roots of this equation that will yield solutions. However, this problem is not about finding the actual roots, it's about finding all the possibilities and for this we need another theorem, which is called the **Rational Root Theorem**.

The Rational Root Theorem says that for any rational equation, all of the possible roots can be found by dividing each of the factors of the constant term (often labeled as "p") by each of the factors of the leading coefficient (often labeled as "q"). In most cases, you will see this written in a way where the factors of the constant term are labeled as p and the factors of the leading coefficient are labeled q

and you must list all of the possible combinations of $\frac{p}{q}$.

To begin, list all the factors of the leading coefficient (in this case 4) and all the factors of the constant term (in this case -12). It's important to note that the negative and positive don't really matter when you identify the number, because we will need to look at both the positive and the negative of each factor.

q = Factors of 12 (same as the factors of -12): $\pm1, \pm2, \pm3, \pm4, \pm6, \pm12$

p = Factors of 4: $\pm1, \pm2, \pm4$

Now that we've listed all the factors, we write all of $\frac{p}{q}$. It's typically easiest to do this in an organized manner so that you don't miss any possible roots. The easiest way is to write, for a single value of *p*, each of the fractions $\frac{p}{q}$. When you finish with one value of *p*, you can go on to the next.

$$\pm\frac{1}{1},\pm\frac{1}{2},\pm\frac{1}{4},\pm\frac{2}{1},\pm\frac{2}{2},\pm\frac{2}{4},\pm\frac{3}{1},\pm\frac{3}{2},\pm\frac{3}{4},\pm\frac{4}{1},\pm\frac{4}{2},\pm\frac{4}{4},\pm\frac{6}{1},\pm\frac{6}{2},\pm\frac{6}{4},\pm\frac{12}{1},\pm\frac{12}{2},\pm\frac{12}{4}$$

Now that you have a complete list, you can simplify and remove any duplicates. Simplifying gives you the list:

$$\pm\frac{1}{1},\pm\frac{1}{2},\pm\frac{1}{4},\pm2,\pm1,\pm\frac{1}{2},\pm3,\pm\frac{3}{2},\pm\frac{3}{4},\pm4,\pm2,\pm1,\pm6,\pm3,\pm\frac{3}{2},\pm12,\pm6,\pm3$$

And finally, if you remove duplicates, you're left with the list:

$$\pm1,\pm2,\pm3,\pm4,\pm6,\pm12,\pm\frac{1}{2},\pm\frac{1}{4},\pm\frac{3}{2},\pm\frac{3}{4}$$

Remember that these are only the possible roots of the polynomial $4x^5+3x^4-2x^2+6x-12=0$. You would need to use division and/or factoring to determine which of these numbers is a root of the polynomial. Also, remember that because this is a fifth-degree polynomial (the largest exponent is a 5) this polynomial will have five roots (counting repeated and complex roots).

TRY IT!

1. Find all the zeros in $P(x)=3x^4-2x^3-x^2-12x-4$

2. Find all the zeros in $F(x)=x^4-16$

3. Find all the zeros in $H(x)=x^3-2x^2+2x$

Answers

1. **Note**: Solutions are known as zeros, roots, or *x*-intercepts. Remember that *x*-intercepts are the points where y = 0 and the line intersects the x-axis.

   Using the Rational Zeros or Roots theorem from Section 3.3, we obtain the following list of possible rational zeros: $\pm1, \pm2, \pm4, \pm\frac{1}{3}, \pm\frac{2}{3}, \pm\frac{4}{3}$.

Checking these using synthetic division or long division, we find that 2 and -⅓ are zeros, and we get the following factorization:

$$P(x) = 3x^4 - 2x^3 - x^2 - 12x - 4$$
$$= (x-2)(3x^3 + 4x^2 + 7x + 2)$$
$$= (x-2)(x+\frac{1}{3})(3x^2 + 3x + 6)$$
$$= 3(x-2)(x+\frac{1}{3})(x^2 + x + 2)$$

Using the quadratic formula, we arrive at the following list of zeros for P(x):

$$2, -\frac{1}{3}, \frac{-1 \pm \sqrt{-7}}{2}$$

Note that $\frac{-1 \pm \sqrt{-7}}{2}$ is an imaginary zero of this polynomial, while 2 and $-\frac{1}{3}$ are real zeros. $\frac{-1 \pm \sqrt{-7}}{2}$ is more commonly written as $\frac{-1 \pm i\sqrt{7}}{2}$, where *i*, represents $\sqrt{-1}$ (the imaginary unit).

2. $F(x) = x^4 - 16$

To find the zeros it is easiest to notice that this is the difference of two squares, which can be factored as such:

$F(x) = x^4 - 16$
$= (x^2 - 4)(x^2 + 4)$

To find the zeros, we solve the following equation:

$(x^2 - 4)(x^2 + 4) = 0$

$x^2 - 4 = 0$
$(x + 2)(x - 2) = 0$
$x = 2, -2$

$x^2 + 4 = 0$
$x^2 = -4$
$x = \pm\sqrt{-4}$

So, the four zeros are, ±2, $\pm\sqrt{-4}$. $\pm\sqrt{-4}$ is more commonly written as $\pm 2i$, where *i*, represents $\sqrt{-1}$ (the imaginary unit). In this example, there are two real (± 2) and two imaginary ($\pm 2i$) zeros.

3. $H(x) = x^3 - 2x^2 + 2x$

For this problem, we will begin by factoring out the *x*.

$H(x) = x^3 - 2x^2 + 2x$
$= x(x^2 - 2x + 2)$

From here, we know x = 0 is one root. The other root will be discovered using the quadratic formula.

$x^2 - 2x + 2$

$$x = \frac{2 \pm \sqrt{(-2)^2 - 4(1)(2)}}{2(1)}$$

$$= \frac{2 \pm \sqrt{4 - 8}}{2}$$

$$= \frac{2 \pm \sqrt{-4}}{2}$$

Thus, the roots are 0 (a real root), and $\frac{2 \pm \sqrt{-4}}{2}$ (two imaginary roots). Note that $\frac{2\pm\sqrt{-4}}{2}$ can be simplified as

$$\frac{2\pm\sqrt{4}\sqrt{-1}}{2}=\frac{2\pm i\sqrt{4}}{2}=\frac{2\pm 2i}{2}=1\pm i$$.

Exercise 1: Absolute Value

THINK: absolute value represents distance from zero. Can a distance be negative? The answer is no, so the absolute value will make all values positive.

Solve $|4x-3|=13$

When you solve an absolute value equation, it's important to remember that the absolute value is defined as the distance a number is from 0 on a number line. While an absolute value will always result in a positive number (because distance is always positive), that number can be located either to the left of 0 or to the right of 0 and still be the same distance from 0 on the number line. In other words, absolute value equations can have two solutions.

To find the solution to an absolute value equation, we must work two problems—one where the answer is positive (just like it is written above) and one where the answer is negative:

$4x-3=13$		$4x-3=-13$
$4x-3+3=13+3$		$4x-3+3=-13+3$
$4x=16$		$4x=-10$
$\frac{4x}{4}=\frac{16}{4}$		$\frac{4x}{4}=\frac{-10}{4}$
$x=4$	or	$x=-\frac{5}{2}$

Exercise 2: Absolute Value Inequalities

Solve $|3x-7|\le 17$

Just like absolute value equations, for absolute value inequalities you must consider what happens on both sides of 0. When you solve an absolute value inequality, you drop the absolute value symbol and solve one inequality exactly as written. The other inequality you must solve is found by switching the inequality symbol around and changing the sign of the number. So, we're going to solve the following two inequalities:

$3x-7\leq 17$ $\qquad\qquad$ $3x-7\geq -17$

Before we go any further, it's important to understand one other important difference between inequalities and equations with absolute values. When solving inequalities, you must determine whether you are looking for the set of all numbers that satisfy either of the two inequalities or the set of numbers that satisfy both inequalities. You determine this by looking at the direction of the inequality symbol in the original problem. If the symbol is a "less than" symbol (like ours is) then you are looking for the values that satisfy both inequalities, so you will join the two answers with an "AND" statement. If your original symbol was a "greater than" symbol, you would join the two inequalities with an "OR" statement.

$3x-7\leq 17$ AND $3x-7\leq 17$

$3x-7+7\leq 17+7$ AND $3x-7+7\geq -17+7$

$3x\geq -10$ AND $3x\geq -10$

$\frac{3x}{3}\leq\frac{24}{3}$ AND $\frac{3x}{3}\geq\frac{-10}{3}$

$x\leq 8$ AND $x\geq -3\frac{1}{3}$

You can write this answer in a variety of forms. The most popular is probably either bracket notation, which would be written as $\left[-3\frac{1}{3},8\right]$ or set builder notation, which would be written as $\{x|-3\frac{1}{3}\leq x\leq 8\}$.

Note: Parentheses are used if it is < or > without an equals sign. You also can leave the answer exactly as it is.

Exercise 1: Radical Equations

THINK: Radicals cannot be negative, unless we are considering imaginary numbers. This is a restriction on the radical function.

Solve $-2+\sqrt{y-3}=5$

To solve an equation involving one radical, you need to get the radical on one side of the equals sign by itself and everything else on the other side of the equation. Let's do that step first.

$-2+2+\sqrt{y-3}=5+2$ ←Add 2 to both sides.

$\sqrt{y-3}=7$ ←Simplify.

Now that we have the radical on one side, we need to square both sides.

$\left(\sqrt{y-3}\right)^2=7^2$ ←Square both sides.

$y-3=49$ ←When you square a radical, the radical is cancelled.

$y=52$ ←Solve for *y* by adding 3 to both sides.

Since we squared both sides of the equation, it's very important to check the answer, because squaring both sides of an equation, while a valid method to use when solving a problem, can introduce an extra answer called an **extraneous** root. So, let's quickly check 52.

$-2+\sqrt{52-3}=5$

$-2+\sqrt{49}=5$

$-2+7=5$

$5=5$

So y = 52 is a valid solution of this equation.

Exercise 2: Solving a Radical Equation with Two Radicals

Solve $\sqrt{x+2}+\sqrt{3x+4}=2$

When solving a radical equation with two radicals, you want to get one on one side of the equals sign and the other on the other side of the equals sign, so let's start there. It doesn't matter which of the radicals you move to the other side of the equals sign so let's move the first one.

$\sqrt{3x+4}=2-\sqrt{x+2}$

Now we square both sides. This calculation gets a bit tricky because you must remember to use FOIL on the right-hand side of the equation where you have two terms.

$\left(\sqrt{3x+4}\right)^2=\left(2-\sqrt{x+2}\right)^2$ ←Square both sides.

$3x+4=\left(2-\sqrt{x+2}\right)\left(2-\sqrt{x+2}\right)$ ←Multiply. It is probably easiest to do the multiplication on the right side of the equation by writing out the perfect square like we've done here. Now we multiply using FOIL. Remember FOIL means to multiply the first term times the first term, then the outside term times the outside term, then the inside term times the inside term, then the last term times the last term.

$3x+4=4-2\sqrt{x+2}-2\sqrt{x+2}+x+2$

←FOIL and combine like terms.

$3x+4=6+x-4\sqrt{x+2}$

Now that we've gotten rid of one of the radicals, we solve this problem just like we solved the previous problem by getting everything but the radical on one side, squaring both sides, and solving. Again, it will be really important to check our answer to make sure we didn't introduce any extraneous roots when we solved the problem.

$3x+4=6+x-4\sqrt{x+2}$

$3x+4-6-x=6-6+x-x-4\sqrt{x+2}$ ←Subtract the 6 and the x from both sides.

$2x-2=-4\sqrt{x+2}$

Now we square both sides:

$(2x-2)^2=\left(-4\sqrt{x+2}\right)^2$

$(2x-2)(2x-2)=16(x+2)$

$4x^2-4x-4x+4=16x+32$ ←Multiply the left using FOIL and distribute the 16 through the parentheses.

$4x^2-8x+4=16x+32$ ←Combine like terms.

$4x^2-8x-16x+4-32=16x-16x+32-32$ ←Set one side equal to 0.

$4x^2-24x-28=0$ ←Combine like terms.

Now you need to factor this equation.

$4x^2-24x-28=0$

$4\left(x^2-6x-7\right)=0$ ←Factor out the greatest common factor, 4.

$4(x-7)(x+1)=0$ ←Factor.

So, $x-7=0$ or $x+1=0$. This means that $x=7$ or $x=-1$. We need to check both of these in our original problem.

Check $x=7$

$\sqrt{x+2}+\sqrt{3x+4}=2$

$\sqrt{7+2} + \sqrt{3(7)+4} = 2$

$\sqrt{9} + \sqrt{25} = 2$

$3 + 5 = 2$

$8 \neq 2$

Since *x* = 7 gives a false answer when you check it, this is an extraneous root and should not be given as a solution to this problem. Let's check the other answer.

Check *x* = -1

$\sqrt{x+2} + \sqrt{3x+4} = 2$

$\sqrt{(-1)+2} + \sqrt{3(-1)+4} = 2$

$\sqrt{1} + \sqrt{1} = 2$

$1 + 1 = 2$

$2 = 2$

Since *x* = -1 gives a true answer when you check it, this is a solution of the problem.

Exercise 1: Interest Problems

You have $2,000 to invest in the bank. The bank is offering you 6% interest (compounded continuously) if you leave the money in the bank for 5 years. If you take their offer, how much money will you have after 5 years?

There is a relatively simple formula you can use to solve continuous compounding problems. It is $A(t) = A_0e^{kt}$ where *A*(*t*) is the amount you will have after *t* years; A_0 is the original amount you invested; and *k* is the interest rate. It's important to remember that *e* is not a variable. The value *e* is a constant, much like π. You normally can find *e* on your calculator, but if you are unsure of where to find this on your calculator you can use the fact that $e \approx 2.718$ to solve the problem. So, now that we know the values of all of our variables, let's substitute everything into the formula.

$A(t) = A_0e^{kt}$

$A(5) = 2000e^{(0.06)(5)}$ ←Substitute the values into the formula.

$A(5) = 2000e^{0.3}$ ←Simplify (0.06)(5).

$A(5) = 2000(1.350)$ ←Simplify $e^{0.3}$.

$A(5) = 2700$

So, after 5 years, you will have $2,700 in your account.

Exercise 2: Domain and Range

Find the domain and range of the function $f(x) = \sqrt{x-1}$.

The **domain** of a function are all the possible values that can be assigned to *x* in the function.

The **range** of a function are all the possible values that can result for *y,* or, in this case, *f(x).*

Let's think about all the values we can put into this equation for *x*. Technically we could substitute in any value for *x*, but something happens if we substitute in values less than 1. If you think about that for a second, you'll realize that if we substitute any value less than 1 for *x,* we get a negative under the radical, which is not a real number, so we need to limit the domain to only those values greater than or equal to 1.Therefore, our domain is $x \geq 1$.

Now, let's think about what this means for the range. If the smallest value we can substitute into this equation for *x* is 1, then when we do that we get $y = 0$. Therefore, the range must be only those values greater than or equal to 0, so the range is $y \geq 0$.

So, the domain is D = $\{x \geq 1\}$ and the range is R = $\{y \geq 0\}$.

Exercise 3: Function Notation

If $f(x) = 3x - 2$ and $g(x) = \frac{x}{3}$ find the following:

a) $(f \circ g)(x)$

b) $(g \circ f)(x)$

c) $f^{-1}(x)$

You can perform a variety of operations on functions, but as you move forward in your study of mathematics, two of them are very important to understand – the composition (which is what we are doing in examples a and b below) and the inverse (which is what we are finding in Example c). First, let's look at the two composition problems.

a) $f \circ g(x)$

Another way of writing *f* of *g* of *x* is to write it like this: $f(g(x))$. This means we take the value of *g*(*x*) and substitute it in for *x* in *f*. To do this we will substitute $g(x) = \frac{x}{3}$ into *f*(*x*).

So, $f \circ g(x)$ = $f(g(x))$= $f(\frac{x}{3}) = 3\left(\frac{x}{3}\right) - 2$

If we simplify this, we get *x* – 2 because the threes cancel each other out.

Now, let's look at Example b:

b) $g \circ f(x)$

This is the same as $g(f(x)) = g(3x-2) = \frac{3x-2}{3}$

Finally, let's look at Example c:

c) $f^{-1}(x)$

To find the inverse of a function, you interchange the *x* and the *y* and then you solve for *y* again.

So, our original function was $f(x) = 3x - 2$

$f(x) = 3x - 2$ is the same as $y = 3x - 2$

Now, we interchange the *x* and the *y* so this gives us $x = 3y - 2$

Now, solve this for *y*:

$x = 3y - 2$

$x + 2 = 3y$ ←Add 2 to both sides.

$\frac{x+2}{3} = \frac{3y}{3}$ ←Divide both sides by 3.

$\frac{x+2}{3} = y$ ←Simplify.

Terms to Know

Absolute Value - Distance from zero; always a positive value.

Conic Section - A conic section is a curve obtained by dissecting a cone (parabola, hyperbola, ellipse, circle).

Determinant - The determinant $\begin{vmatrix} a_1 & b_1 \\ a_2 & b_2 \end{vmatrix}$ is defined to mean $a_1b_2 - a_2b_1$.

Domain - The set of all inputs in a function.

Elimination - A strategy for solving systems of equations that includes elimination of one variable.

Equation - A mathematical statement that two quantities are equal.

Expression - A mathematical statement that does not include an equality or equals sign.

Fundamental Theorem of Algebra - States an equation will have exactly as many roots as its highest degree.

Graphing - A strategy for solving systems of equations that includes graphing the two lines and identifying the point of intersection.

Matrices - A strategy for solving systems of equations that includes row elimination of matrices.

Range - The set of all outputs in a function.

Substitution - A strategy for solving systems of equations that include solving for one variable and using that solution to find the other variable.

Systems of Equations - Two linear equations that are meant to be solved simultaneously; the solution is the intersection of the two points.

Geometry

Geometry is the study of measurement as well as the study of logic. In most geometry courses, students are introduced to the concept of a proof, which is a method of following a logical argument by making step by step progress toward a goal. This study typically begins with a brief introduction to logical arguments and then progresses through a lengthy study of methods and formats of proof.

Students in geometry classes study the properties of objects in the coordinate plane (on an x and y axis) as well as objects in general. The shapes that we are generally concerned with in geometry are angles, triangles, quadrilaterals (shapes with four sides), **polygons** (shapes in n number of sides), and circles. Students study these objects using a number of different methods including actual measurement, constructions using a compass and straightedge, and computer software.

Geometry and the study of proof and logic make use of postulates and theorems. **Postulates** are statements that are accepted as true without the need for proof. **Theorems** are statements that have been proved using postulates, definitions, and previously proved theorems.

Finding the Distance Between Two Points on the Coordinate Plane

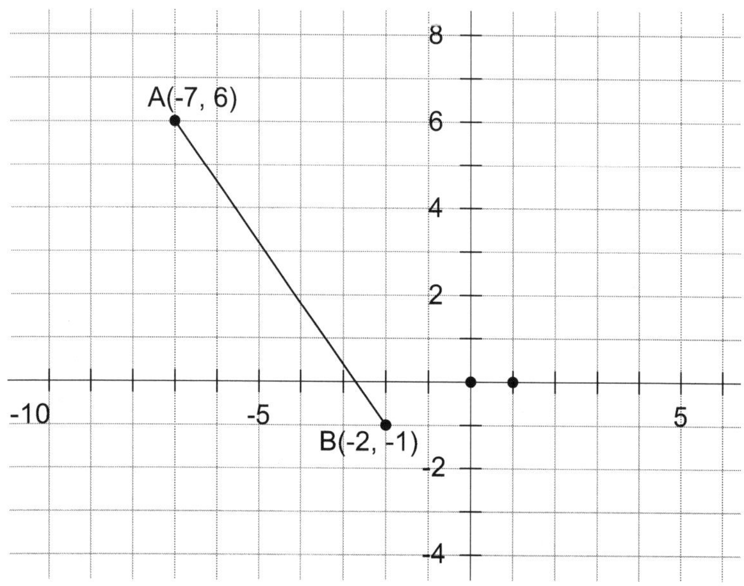

Example 1: Find the distance between the points shown on the graph below.

The two points are A (-7,6) and B (-2,-1). To find the distance between these points, use the distance formula, which is

$$d = \sqrt{(x_1 - x_2)^2 + (y_1 - y_2)^2}\ .$$

When we substitute the values for these two points into the formula, we get:

$$d = \sqrt{(-7 - -2)^2 + (6 - -1)^2}$$

$$d = \sqrt{(-5)^2 + (7)^2}$$ ← Simplify.

$$d = \sqrt{25 + 49}$$

$\sqrt{74}$ units

These two points are $\sqrt{74}$ units apart.

Exercise 1

Find the distance between the two points shown and then find the midpoint between the two points.

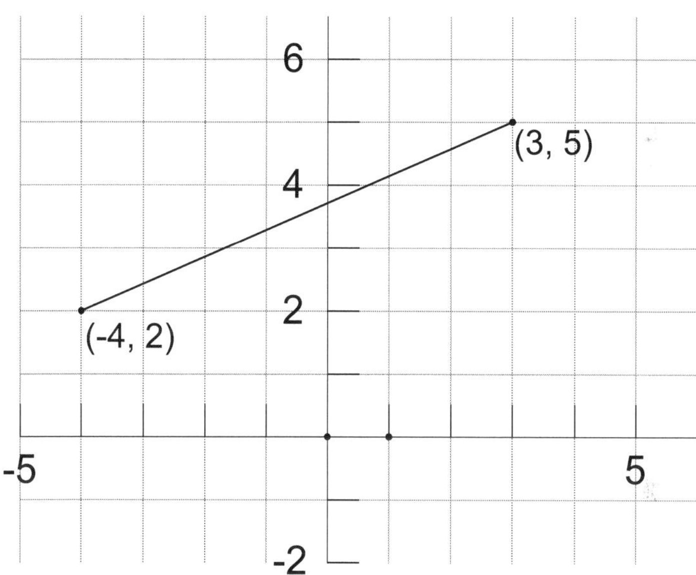

We use the **distance formula** to find the distance between two points. The distance formula is $d = \sqrt{(x_1 - x_2)^2 + (y_1 - y_2)^2}$.

We'll substitute the points into this formula and simplify.

$d = \sqrt{(-4-3)^2 + (2-5)^2}$

$d = \sqrt{(-7)^2 + (-3)^2}$

$d = \sqrt{49+9}$

$d = \sqrt{58}$

TRY IT!

1. Find the distance between the following points:

   a) A (-5,2) and B (-3, 1)

   b) B (-3, 1) and C (3, -2)

2. True or False: B is the midpoint of $\overline{AC}$.

Use the coordinates above to show that AB=BC, making B the midpoint of $\overline{AC}$.

Answers

$$AB = \sqrt{(-5-(-3))^2 + (2-1)^2} = \sqrt{(-2)^2 + 1^2} = \sqrt{5}$$
$$BC = \sqrt{(-3-3)^2 + (1-(-2))^2} = \sqrt{(-6)^2 + 3^2} = \sqrt{45} = 3\sqrt{5}$$
$$AC \neq BC$$

So, B is not the midpoint of the line AC. This makes the statement in number 2 false. This is known as a coordinate proof, as we are using coordinates to determine if the statement is true or false.

To find the midpoint of the line segment, we use the midpoint formula and simplify.

$$M = \left(\frac{x_1 + x_2}{2}, \frac{y_1 + y_2}{2}\right)$$
$$M = \left(\frac{-4+3}{2}, \frac{2+5}{2}\right)$$
$$M = \left(-\frac{1}{2}, \frac{7}{2}\right)$$

Example 2

In the figure below, angles 1, 2, and 3 lie on a straight line as do angles 4, 5, and 6. It is common to abbreviate the phrase "the measure of the angle" with the symbols $m\angle$. Suppose we know that $m\angle 1 = x + 20$, $m\angle 2 = x + 40$, and $m\angle 3 = x + 30$. Find the measures of each of the angles 1 – 6.

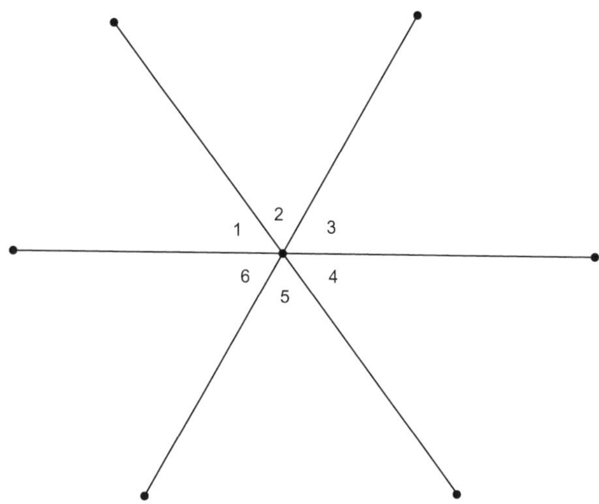

THINK: You may find it easier to write the angle measurements given in the problem on the diagram. Feel free to do that if it helps you visualize how to begin the problem.

You are given the measurements of angles 1, 2, and 3. These three angles together form a straight line, so the sum of their measurements is 180. **THINK:** All straight angles are equal to 180° and two adjacent angles that add up to be 180° are known as a linear pair.

We use this fact and a little algebra to give us $x+20+x+40+x+30=180$. Simplifying this gives $3x + 90 = 180$. Now, simplify this statement using algebra to solve for x.

$3x + 90 = 180$

$3x + 90 - 90 = 180 - 90$	← Subtract 90 from both sides.
$3x = 90$	← Simplify.
$\frac{3x}{3} = \frac{90}{3}$	← Divide both sides by 3 to solve for x.

$x = 30$

Use the value for x to find the measure of angles 1, 2, and 3.

$m\angle 1 = x + 20 = 30 + 20 = 50^\circ$

$m\angle 2 = x + 40 = 30 + 40 = 70^\circ$

$m\angle 3 = x + 30 = 30 + 30 = 60^\circ$

Now, we can use this information and the fact that angles arranged vertically (the two cross-sections of an X) are equal in measure to find the measures of angles 4, 5, and 6. (**THINK: Vertical angles** are congruent or equal in measure.)

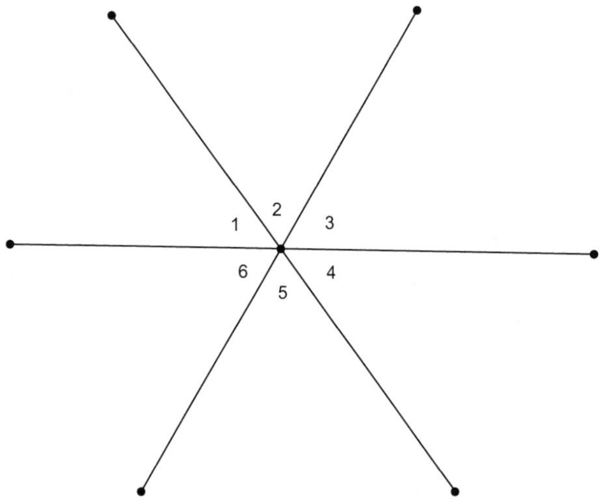

TRY IT!

1. Using the new values for angle one and angle 3, find the measure of all of the angles in the above diagram. Offer an explanation for your final answers.

$m\angle 1 = 35^\circ$

$m\angle 3 = 22^\circ$

$m\angle 2 = ?$

$m\angle 4 = ?$

$m\angle 5 = ?$

$m\angle 6 = ?$

Answers

1.
$$m\angle 1 = 35^\circ$$
$$m\angle 3 = 22^\circ$$
$$m\angle 2 = 123^\circ$$
$$m\angle 4 = 35^\circ$$
$$m\angle 5 = 123^\circ$$
$$m\angle 6 = 22^\circ$$

Explanation:

We are given that $m\angle 1 = 35^\circ$ and $m\angle 3 = 22^\circ$. Recognizing that $\angle 1$, $\angle 2$, and $\angle 3$ form a straight angle, we can write and solve the following equation:

$$m\angle 1 + m\angle 2 + m\angle 3 = 180^\circ$$
$$35^\circ + m\angle 2 + 22^\circ = 180^\circ$$
$$m\angle 2 = 123^\circ$$
$$m\angle 1 = m\angle 4 = 35^\circ$$
$$m\angle 2 = m\angle 5$$
$$m\angle 6 = m\angle 3$$

THINK: Vertical angles are congruent or equal in length.

so:

$$m\angle 1 = 35^\circ$$
$$m\angle 3 = 22^\circ$$
$$m\angle 2 = 123^\circ$$
$$m\angle 4 = 35^\circ$$
$$m\angle 5 = 123^\circ$$
$$m\angle 6 = 22^\circ$$

Example 3: Parallel Lines Cut by a Transversal

In the figure below, *j* || *k* and *l* || *m*. These are parallel lines cut by a non-parallel line or transversal. So, *j* is the transversal of lines *l* and *m*. Find the values of x, y, and z.

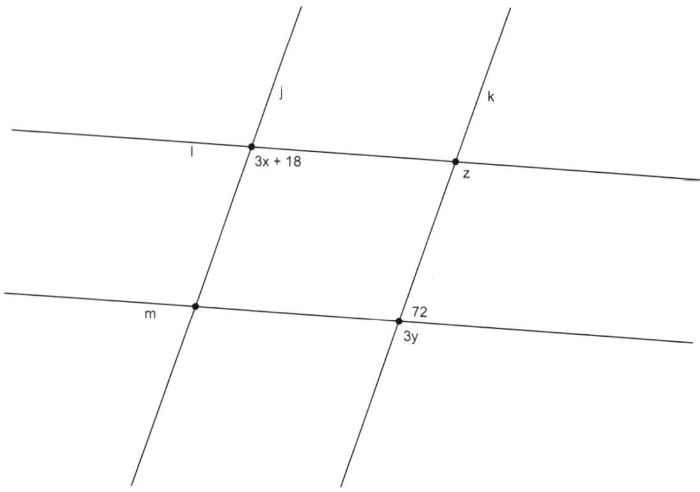

There are several important relationships that exist when parallel lines are cut by another line (transversal). In the example above, two transversals that happen to be parallel to one another are cutting through another set of parallel lines. Let's review those relationships that exist with parallel lines.

Alternate Interior Angles Theorem*:* If two parallel lines are cut by a transversal, then each pair of alternate interior angles is congruent.

Consecutive Interior Angles Theorem*:* If two parallel lines are cut by a transversal, then each pair of consecutive interior angles is supplementary.

Alternate Exterior Angles Theorem*:* If two parallel lines are cut by a transversal, then each pair of alternate exterior angles is congruent.

Let's apply these theorems to this problem. The angles marked z and 72 are consecutive interior angles. Therefore, they are supplementary so $z + 72 = 180$. Subtracting 72 from both sides gives $z = 108$.

We also know that the angle marked $3y$ and the angle marked 72 are located on a straight angle. Therefore the two of them together add up to 180°. So, $3y + 72 = 180$. We solve this as follows:

$3y + 72 = 180$

$3y + 72 - 72 = 180 - 72$ ← Subtract 72 from both sides.

$3y = 108$

$\frac{3y}{3} = \frac{108}{3}$ ← Divide both sides by 3.

$y = 36$

Finally, we know that the angle marked $3y$ is equal to $3 \cdot 36$ (substituting in the value we just found for y), which is 108.

This angle is equal to $3x + 18$ so we can set this equal to 108.

$3x + 18 = 108$

$3x + 18 - 18 = 108 - 18$ ← Subtract 18 from both sides.

$3x = 90$

$\frac{3}{3}x = \frac{90}{3}$ ← Divide both sides by 3.

$x = 30$

Exercise 1

In the figure below, line *j* is parallel to line *k*. Use that fact to find the value of *x*.

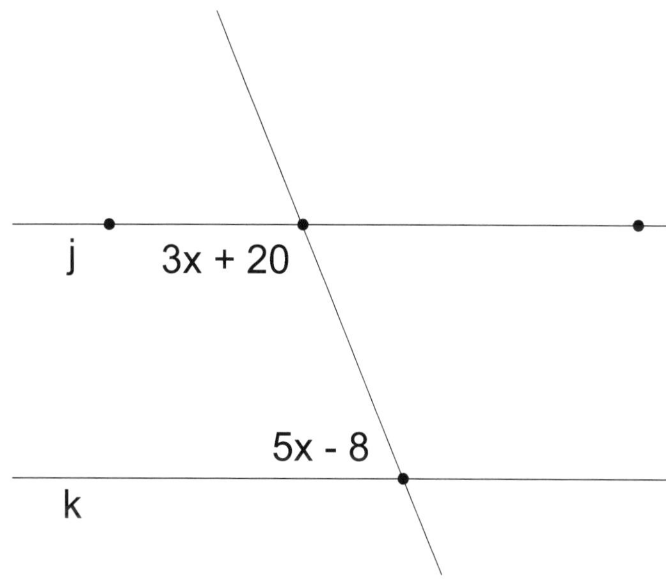

These two angles are consecutive interior angles. From the theorem we saw above, we know that these two angles are supplementary, which means the sum of their measures is 180 degrees. To solve this problem, we'll add these two values and set them equal to 180.

$3x + 20 + 5x - 8 = 180$

$8x + 12 = 180$

$8x = 168$

$x = 21$

Exercise 2

In the figure below, find the measures of the angles marked *x* and *y*.

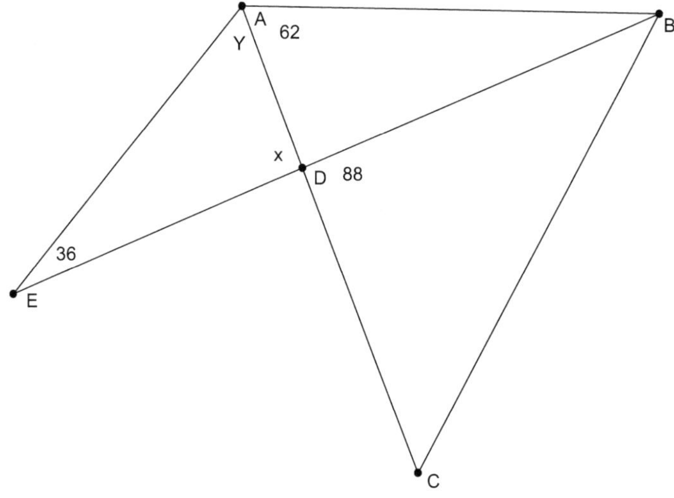

The angle marked *x* (angle ADE) and angle BDC are vertical angles, which means they are congruent. Therefore, $x = 88$.

Once we know that *x* is equal to 88, we can find *y* by using the fact that the sum of the measures of the angles of a triangle is 180. Therefore, $36 + 88 + y = 180$.

$36 + 88 + y = 180$

$124 + y = 180$

$124 - 124 + y = 180 - 124$

$y = 56$

TRY IT!

1. Find the measures of x, y, and z.

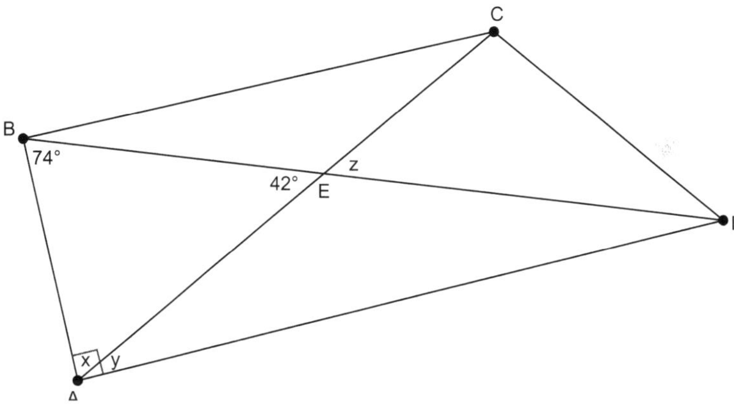

2. Find the measures of x and y.

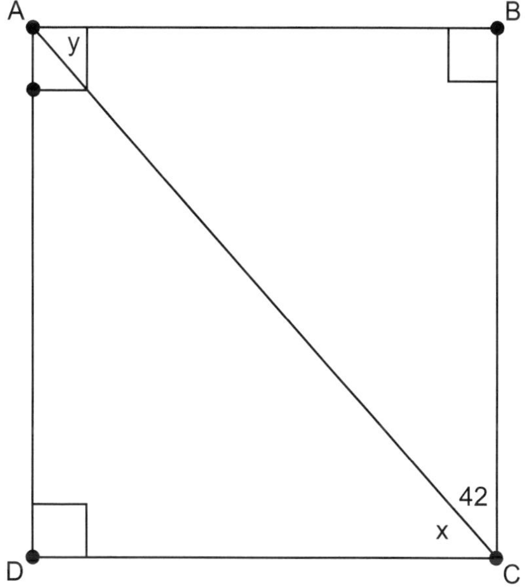

Answers

1. $x = 64^\circ, y = 26^\circ$, and $z = 42^\circ$.

$\angle$AEB and $\angle$CED are vertical angles, and therefore they are congruent. This means that *z* = 42°. To find *x*, we need to remember that the sum of the measures of the angles of a triangle is 180°. Working with ΔABE, we can then write and solve the equation:

$$x + 42 + 74 = 180$$
$$x + 116 = 180$$
$$x = 64^\circ$$

To find *y*, we note that $\angle$BAD is marked as a right angle. Therefore, *x* + *y* = 90, and we can substitute the value of *x* we found above to find *y*:

$$64 + y = 90$$
$$y = 26^\circ$$

2. $x = 48^\circ, y = 48^\circ$

To answer this question, we first recognize that $\angle$BCD also must be a right angle. We know this because the sum of the measures of the angles of a quadrilateral is 360°, and there are already three right angles marked in the diagram. These three angles account for 270°, which leaves 90° leftover for $\angle$BCD. Another way to reach the same conclusion is to recognize that ABCD must be a rectangle, and therefore all four angles are right angles (which each measure 90°). We can then find *x* by solving the equation:

$$x + 42 = 90$$
$$x = 48^\circ$$

To find *y*, we make use of the Alternate Interior Angles Theorem. $\overline{AB} \parallel \overline{CD}$, which makes $\angle$BAC and $\angle$ACD alternate interior angles and therefore congruent. Since *x* is 48°, *y* is also 48°.

Example 4

Polygon *ABDEC* is similar to polygon *J'F'G'H'I'*. Find the scale factor of polygon *ABDEC* to polygon *J'F'G'H'I'*. Then find the values of *x* and *y*.

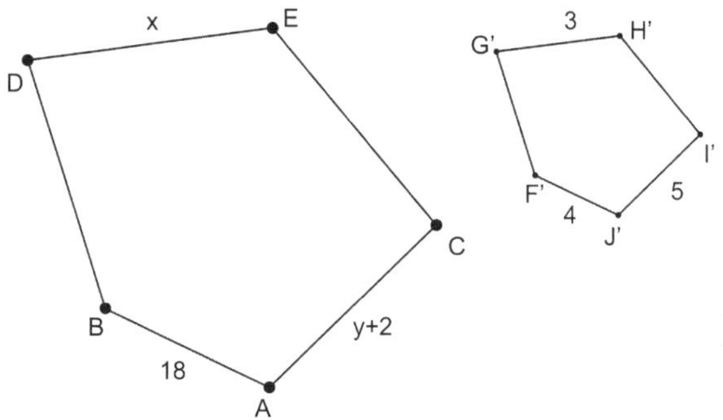

The two figures are **similar**. Two figures are similar if they have the same shape and if the length of each set of corresponding sides has the same scale factor. The **scale factor** is the ratio of the lengths of two corresponding sides. We know that side *AB* corresponds to side *J'F'*. Therefore the scale factor is $\frac{18}{4}=\frac{9}{2}$. We'll use this ratio to set up proportions to find the values of *x* and *y*. To find *x*:

$\frac{x}{3}=\frac{9}{2}$	← Set up a ratio of the sides equal to the scale factor.
$2x=27$	← Cross-multiply.
$x=\frac{27}{2}=13.5$	←Divide both sides by 2.

Now we'll use the same scale factor to find *y*:

$\frac{y+2}{5}=\frac{9}{2}$	← Set up a ratio of the sides equal to the scale factor.
$2(y+2)=45$	← Cross-multiply.
$2y+4=45$	← Distribute.

$$
\begin{aligned}
2y + 4 &= 45 \\
-4 \quad & -4 \\
\hline
2y \quad &= 41
\end{aligned}
$$

← Subtract 4 from both sides. Simplify.

$$\frac{2y}{2} = \frac{41}{2}$$

← Divide both sides by 2.

$$y = 20.5$$

← Simplify.

Exercise 6

In the figure below, triangle ABC is similar to triangle AB'C'. Find the values of *x* and *y*.

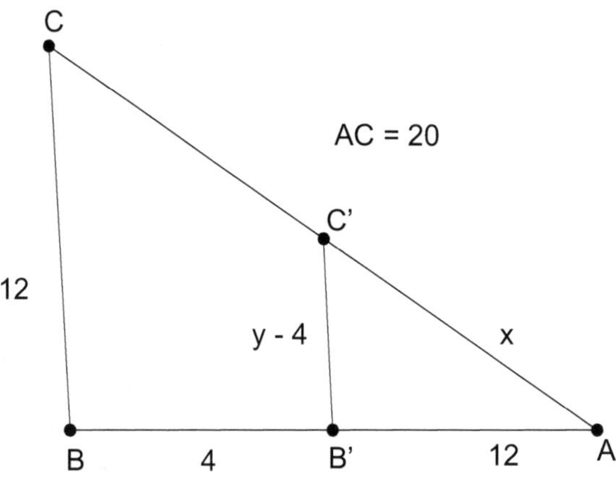

To work with similar triangles, begin by finding the scale factor between triangle ABC and triangle AB'C'. You can do that using the ratio of side AB' to side AB. This ratio is: $\frac{12}{12+4} = \frac{12}{16} = \frac{3}{4}$

We'll now use this ratio to find *x* and *y*. It's important to always place the values in the same places in each ratio. Since we found the scaling factor in terms of the small triangle related to the large triangle (notice the smaller number is on top of the ratio), we need to do the same thing in the rest of our ratios. So, to find *x* we need to set up a ratio between AC' and AC. That ratio is $\frac{x}{20}$. We set that equal to the ratio for our scale factor and then cross-multiply and solve for *x*.

$\frac{x}{20} = \frac{3}{4}$	← Write the proportion.
$4x = 60$	← Cross-multiply.
$x = 15$	← Divide both sides by 4.

Now, we use the same process to find *y*. The ratio for *y* is the ratio of side B'C' to side BC. This is the ratio $\frac{y-4}{12}$. Use the same process as above to solve for *y*.

$\frac{y-4}{12} = \frac{3}{4}$	← Write the proportion.
$4(y - 4) = 36$	← Cross-multiply.
$4y - 16 = 36$	← Distribute the 4 through the parentheses.
$4y - 16 + 16 = 36 + 16$	← Add 16 to both sides.
$4y = 52$	
$y = 13$	← Divide both sides by 4 to solve for *y*.

TRY IT!

1. Find *x*.

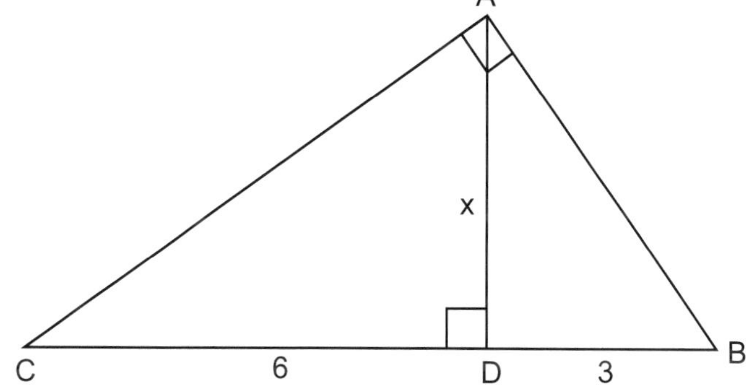

2. Find *y*.

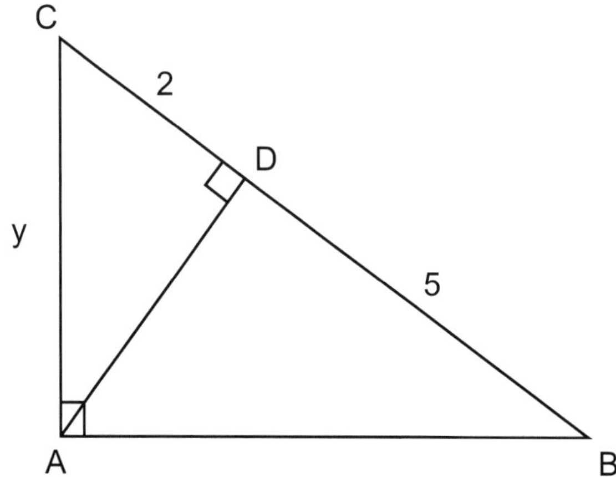

Answers

1. $x = 3\sqrt{2}$. This type of problem is called an "altitude on hypotenuse," since $\overline{AD}$ is an altitude drawn to hypotenuse $\overline{BC}$. Therefore, ΔADC ~ ΔBDA, and we can write the proportion:

$$\frac{AD}{BD} = \frac{DC}{AD}$$

Substituting the values from the diagram, we can solve for *x* as follows:

$\frac{x}{3} = \frac{6}{x}$ ←Substitute.

$x^2 = 18$ ←Cross-multiply.

$x = \pm\sqrt{18}$ ←Take the square root of both sides.

$x = 3\sqrt{2}$ ←Simplify the radical and discard the negative solution.

2. $y = \sqrt{14}$. This is another example of an altitude on hypotenuse problem, since $\overline{AD}$ is an altitude drawn to hypotenuse $\overline{BC}$. Therefore, ΔADC ~ ΔBAC, and we can write the proportion:

$$\frac{AC}{BC} = \frac{DC}{AC}$$

Substituting the values from the diagram, we can solve for *y* as follows:

$\frac{y}{5+2} = \frac{2}{y}$	←Substitute.
$y^2 = 14$	←Cross-multiply.
$y = \pm\sqrt{14}$	← Take the square root of both sides.
$y = \sqrt{14}$	← Discard the negative solution.

Example 5: Pythagorean Theorem and Side Lengths of Right Triangles

In Circle C, the radius has a length of 6 and line segment AB is a tangent of the circle at Point B. Find the value of *x*.

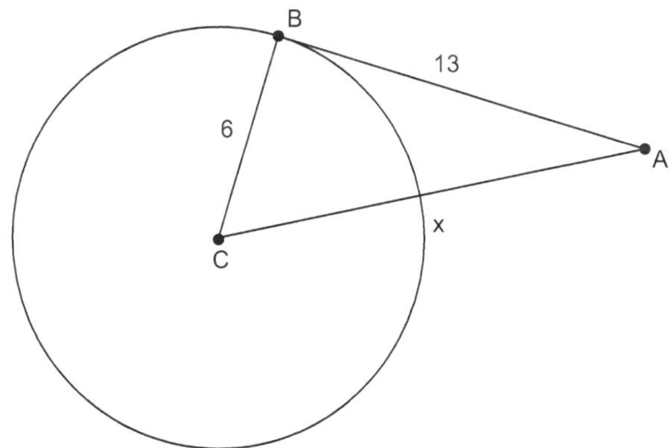

There is a theorem that says if a line is tangent to a circle, then it is perpendicular (at a right angle) to the radius drawn to the point of tangency. Because of this, we know that triangle *ABC* is a right triangle.

All right triangles have a special relationship defined by the Pythagorean Theorem. The **Pythagorean Theorem** says that the sum of the squares of the two shorter sides (called legs) is equal to the square of the longest side (called the hypotenuse). The **hypotenuse** is always the side located across from the right angle. In this case, it is labeled *x*. Therefore, we know that

$6^2 + 13^2 = x^2$ ← Pythagorean Theorem

$36 + 169 = x^2$

$205 = x^2$

$\sqrt{205} = \sqrt{x^2}$

$\sqrt{205} = x$

You can leave the answer in this form or use a calculator to find that $\sqrt{205} \approx 14.3$.

Exercise 1

Find the value of *x* in the right triangle *ABC*.

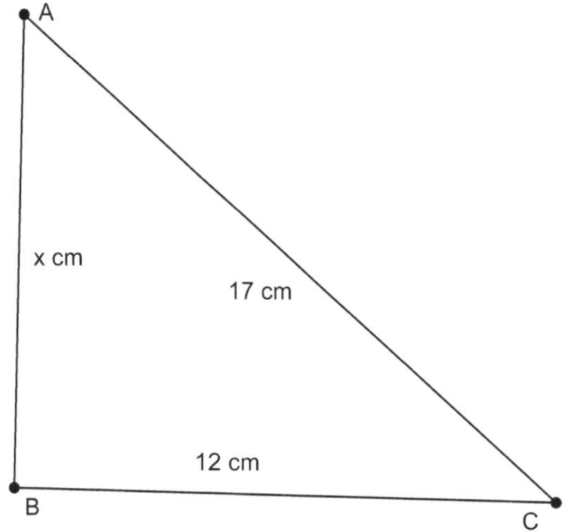

In any right triangle, we can use the Pythagorean Theorem to find the length of a missing side. Sides *AB* and *BC* are the legs. The sum of the squares of these two sides is equal to the square of the hypotenuse.

$x^2 + 12^2 = 17^2$ ← Apply Pythagorean Theorem.

$x^2 + 144 = 289$

$x^2 + 144 - 144 = 289 - 144$

$x^2 = 145$

$x = \sqrt{145}$ cm

Exercise 2

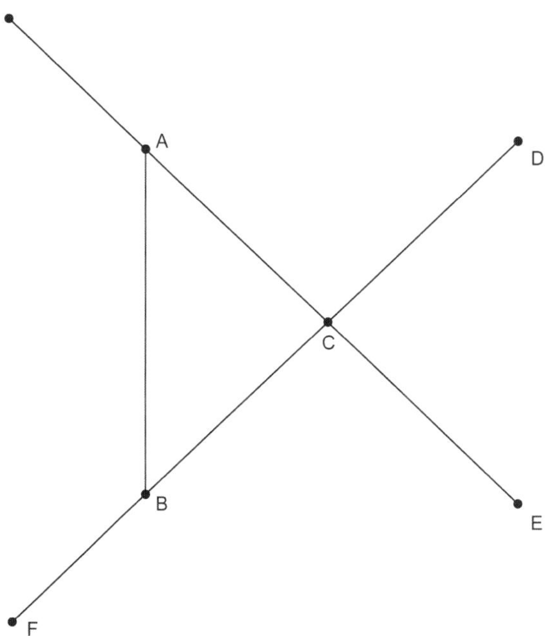

In the figure shown below, C bisects segment AE as well as segment BD. If $BC = 2x + 10$ and $CD = 3x - 4$, find the value of *x*.

THINK: Segment bisectors, like angle bisectors cut a segment or angle in half.

BC = CD	← Definition of midpoint.
$2x + 10 = 3x - 4$	← Definition of midpoint.
$14 = x$	← Simplify using algebra.

Exercise 3

In the figure shown below, the measure of the diagonal is 5 cm. Find the length of a side of the square.

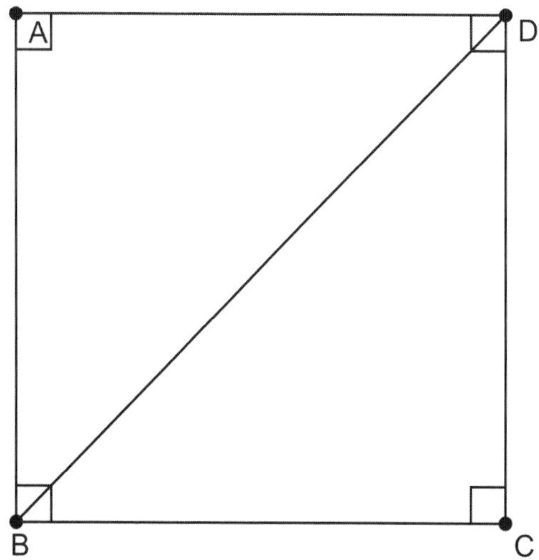

Since the figure is a square, we know that all four sides are equal in length. We also know that the diagonal drawn from one vertex to the opposite vertex forms two right triangles. Since the sides are of equal length, we can let the length of a side be *x*. Therefore, using the Pythagorean Theorem, we know:

$$x^2 + x^2 = 5^2$$

$$2x^2 = 25$$

$$\frac{2x^2}{2} = \frac{25}{2}$$

← The Pythagorean Theorem says the sum of the squares of the legs of a right triangle is equal to the square of the hypotenuse.

$$x = \sqrt{\frac{25}{2}}$$

$$x \approx 3.5$$

Exercise 3

In the triangle shown below, $\angle ACB$ is a right angle and segment *DC* is an altitude drawn to side *AB*. Find the lengths of sides *AC* and *BC* if *BD* = *5* and *AD* = 8.

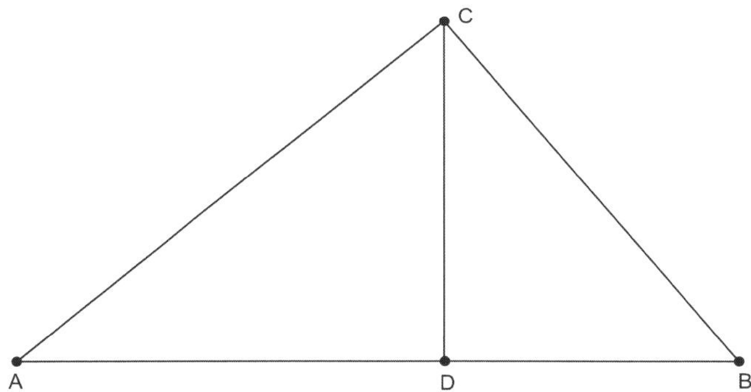

The altitude of a triangle is a line drawn from a point perpendicular to the side opposite to that point. So, since line segment *CD* is an altitude, it meets side *AB* in a right angle at point *D*. We already know the Pythagorean Theorem allows us to find the value of missing sides in a right triangle, but that doesn't give us quite enough information to solve this problem. To solve this problem, we need a few additional theorems.

Theorem*:* If an altitude is drawn from the vertex of the right angle of a right triangle to its hypotenuse, then the two triangles formed are similar to the given triangle and to each other.

Theorem*:* The measure of the altitude drawn from the vertex of the right angle of a right triangle to its hypotenuse is the geometric mean between the measures of the two segments of the hypotenuse.

Theorem*:* If the altitude is drawn to the hypotenuse of a right triangle, then the measure of a leg of the triangle is the geometric mean between the measures of the hypotenuse and the segment of the hypotenuse adjacent to that leg.

These last two theorems both use the concept of a geometric mean. Let's take a look at what that is before we move forward with solving this problem.

The geometric mean between two positive numbers *a* and *b* where $x = \sqrt{ab}$.

We'll use that to solve this problem. We can label our diagram as shown below with the information given in the problem.

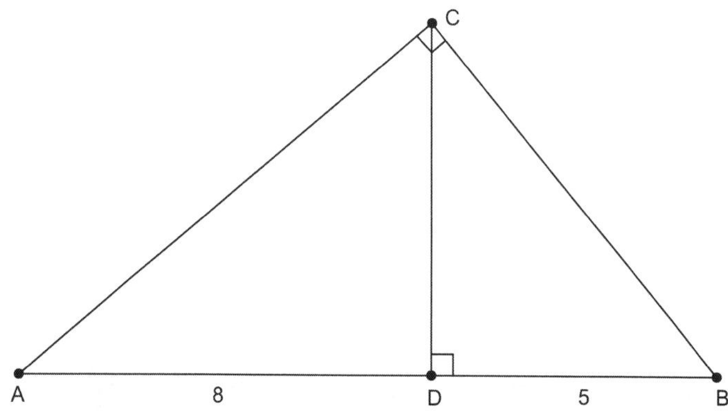

We know from the third theorem above that the measure of a leg of the right triangle is the geometric mean between the measure of the hypotenuse and the segment of the hypotenuse adjacent to that leg. Let's use that information to find the length of *BC* first. The measure of the hypotenuse is 8 + 5 = 13. The measure of the segment of the hypotenuse adjacent to leg *BC* is the length of segment *DB* which is 5. So, we need to find the geometric mean between 13 and 5 which is:

$BC = \sqrt{5 \cdot 13} = \sqrt{65} \approx 8.1$ units

We use the same process to find the length of *AD*. *AD* is going to be the geometric mean between *AD* and *AB*. This is:

$AD = \sqrt{8 \cdot 13} = \sqrt{104} \approx 10.2$ units

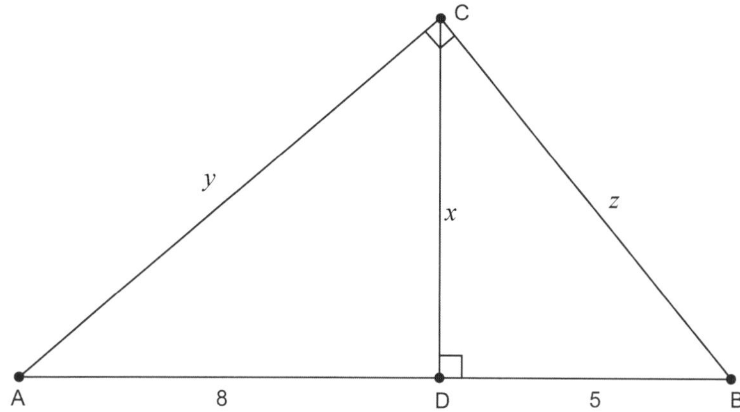

There is another way to work this problem using only the first theorem shown above and the Pythagorean Theorem. In the example shown below, to simplify the notation we're going to let side *CD* = *x*, side *AC* = *y*, and side *BC* = *z*.

In this method, you would find the geometric mean of the two segments of the hypotenuse this way:

$\frac{8}{x} = \frac{x}{5}$	←Find the geometric mean of 8 and 5.
$x^2 = 40$	←Cross-multiply.
$x = \sqrt{40}$	←Take the square root of both sides.
$x = 2\sqrt{10}$	←Simplify (it will be easier to use the $\sqrt{40}$ in the next step though so we'll use that.)

Now you'll use the fact that $8^2 + \sqrt{40}^2 = y^2$ and 8 are two legs of a right triangle and that $x = \sqrt{40}$ and 5 are two legs of a right triangle. You'll use the Pythagorean Theorem two times with these two sets of values to find the lengths of the missing sides. We'll work the two Pythagorean Theorem problems side by side.

$8^2+\sqrt{40}^2=y^2$	←Substitute→	$5^2+\sqrt{40}^2=z^2$
$64+40=y^2$	←Simplify→	$25+40=z^2$
$104=y^2$	←Simplify→	$65=z^2$
$\sqrt{104}=y$	←Take the square root of both sides→	$\sqrt{65}=z$

TRY IT!

1)

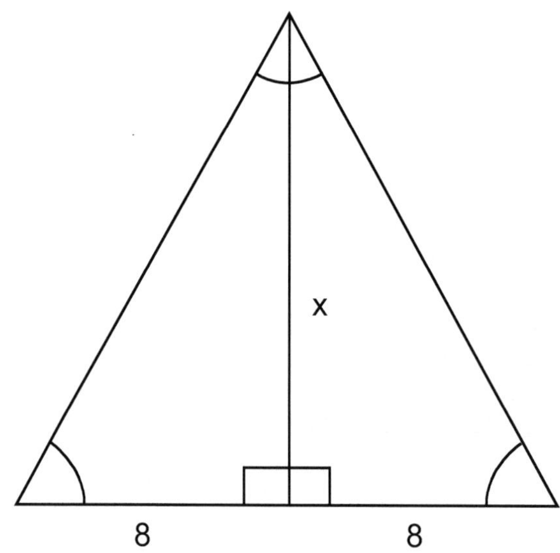

Solution:

This is an equiangular, equilateral triangle, so all sides are congruent and equal to 16. This means we can focus on one of the right triangles.

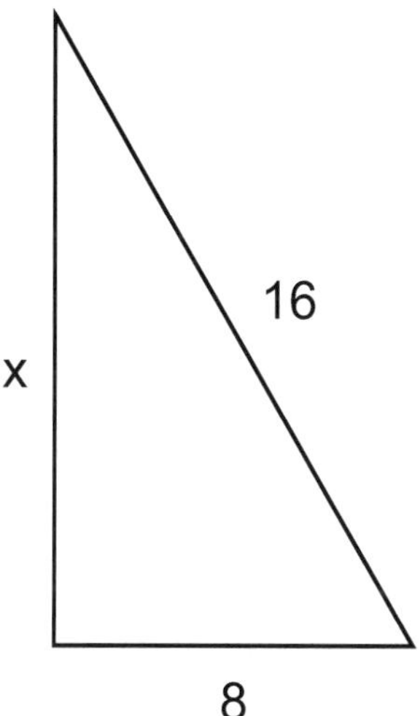

So,

$$x^2 + 8^2 = 16^2$$
$$x^2 + 64 = 256$$
$$x^2 = 192$$
$$x = \sqrt{192} = \sqrt{16 \cdot 4 \cdot 3} = 4 \cdot 2\sqrt{3} = 8\sqrt{3}$$
$$x \approx 13.86$$

2)

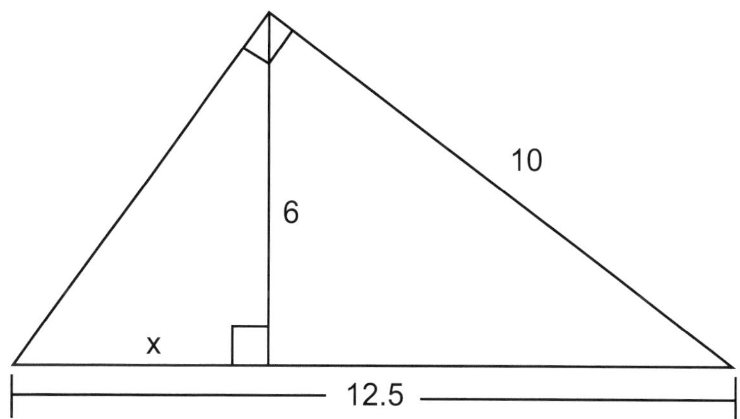

Solution:

This is a multistep problem that requires us to look at this figure from the largest triangle perspective. Namely, we are looking to find the missing side of the largest right triangle with hypotenuse 12.5 and one leg equal to 10. To do this we will call the missing side *y* and solve using the Pythagorean Theorem.

$$x^2 + 10^2 = (12.5)^2$$

$$x^2 + 100 = 156.25$$

$$x^2 = 56.25$$

$$x = 7.5$$

With this side, we can now find the length of *x*, as the hypotenuse of one of the smaller triangles is 7.5, one leg is *x* and the other is 6. To find the value of *x*, we use the Pythagorean Theorem.

$$x^2 + 6^2 = 7.5^2$$

$$x^2 + 36 = 56.25$$

$$x^2 = 20.25$$

$$x = 4.5$$

3)

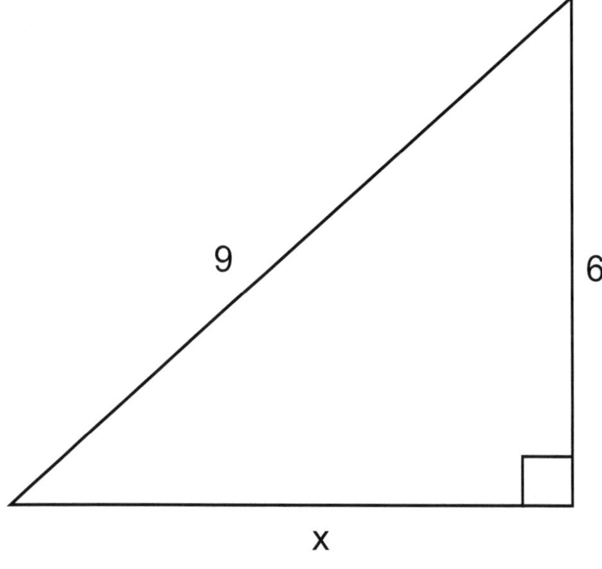

Solution:

$x^2 + 6^2 = 9^2$

$x^2 + 36 = 81$

$x^2 = 45$

$x = \sqrt{45} = \sqrt{9 \cdot 5} = 3\sqrt{5} \approx 6.71$

Properties of Circles

Exercise 1

In Circle C shown below, find the length of arc *XY* if the length of the radius is 6 units.

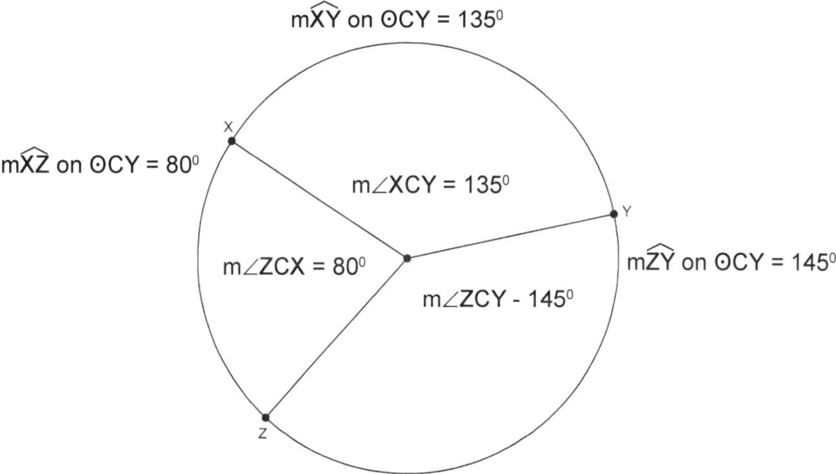

As you can see from the figure, the measure of the central angle (the angle formed by two radii) is equal to the measure of the angle of the arc that the angle includes. To find the arc length (the distance between the two points), we have to find out what portion of the total distance around the circle (or what portion of the circle's circumference) is located between the two points. Arc angle *XY* has a measure of 135°. This is $\frac{135}{360} = \frac{3}{8}$ of the circle's total circumference. We know that the circumference can be found using the formula:

Circumference = $2\pi r$

Circumference = $2\pi(6) = 12\pi \approx 37.70$

Now that we know the circumference, we just need to find $\frac{3}{8}$ of 37.70 so:

$\frac{3}{8} \cdot 37.70 \approx 14.14$ units

This exercise reviewed the process for finding the circumference of a circle. It also showed us that central angles are equal to their intercepted arc.

Exercise 2

In the circle shown below, find the measure of angle *ADC*.

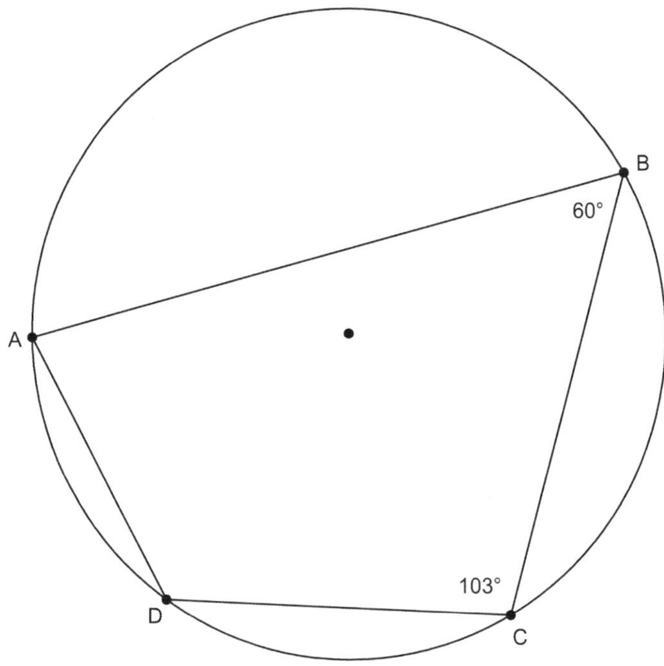

If a quadrilateral is **inscribed** in a circle (all four of its vertices are on the circle), then the angles opposite one another in the quadrilateral are supplementary. Recall that if two angles are supplements, then the sum of their measures is 180 degrees.

Therefore, since we know angles *ABC* and *ADC* are opposite one another in this quadrilateral, the sum of their measures must be 180 degrees. This tells us:

$ADC + 60 = 180$

$ADC = 180 - 60 = 120$ degrees. This exercise modeled an important property of circles and inscribed polygons. Namely, if a quadrilateral is touching the circle in which it is inscribed, the opposite angles of the polygon are **supplementary** or sum to180°.

Finding the sum of the interior angles of a regular polygon.

Exercise 1

First it is important to define a "regular figure" as one with all sides and angles congruent. We are going to practice that concept in this problem.

Find the sum of the measures of the interior angles of a regular decagon.

A decagon is a figure with 10 sides ("deca" means 10; consider the word "decade," which means 10 years). This problem is very easy if we use the Interior Angle Sum Theorem: If a convex polygon has *n* sides then *S,* which is the sum of the measures of its interior angles, can be found using the formula $S = 180(n-2)$.

Since a decagon has 10 sides, we'll let $n = 10$, so $S = 180(10 - 2) = 180(8) = 1440$ degrees.

$$S = 180(n-2)$$

Note: This is a very important formula that should be committed to memory.

Coordinate Proof

A geometric proof is one that uses coordinates to verify or prove a statement. Such as using coordinates to determine if a figure is a specific type of polygon. Remember in order to be a square, all sides of a figure must be congruent. To be a rectangle, opposite sides are congruent. A parallelogram must have opposite sides congruent. So, a rectangle and a square are both parallelograms.

Exercise 1

The coordinates of the vertices of a quadrilateral are at *A*(3, 7), *B* (7, 1), *C*(1, -3), and *D* (-3, 3). Determine if the figure shown is a square, rectangle, parallelogram, or none of the these.

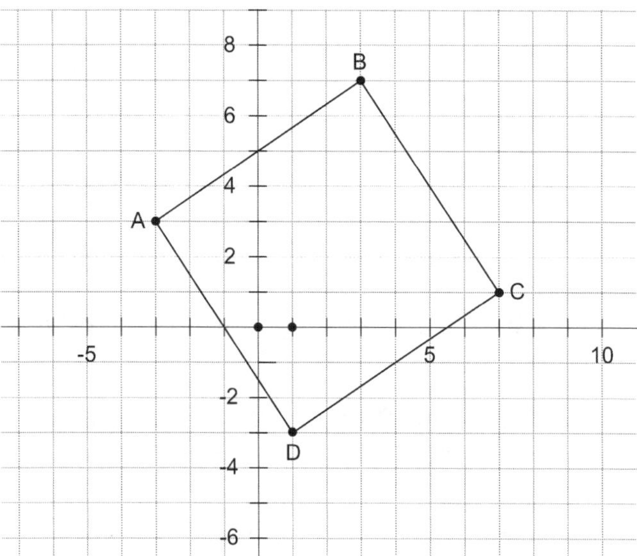

To determine what type of figure this is, we'll need to know the length of each side, which we can find using the distance formula. We also need to know whether the sides meet at right angles, which, if you recall from algebra, we can determine by using the slopes. Remember that if the slopes are opposite reciprocals of one another, then the lines meet at right angles. Let's start with the slopes. We need to find the slopes of all four of the sides. Remember that the slope formula is $m = \frac{y_1 - y_2}{x_1 - x_2}$.

Let's use this to find the slope of each of the sides:

$$m_{AB} = \frac{7-1}{3-7} = -\frac{6}{4} = -\frac{3}{2}$$
$$m_{BC} = \frac{1--3}{7-1} = \frac{4}{6} = \frac{2}{3}$$
$$m_{CD} = \frac{-3-3}{1--3} = -\frac{6}{4} = -\frac{3}{2}$$
$$m_{DA} = \frac{3-7}{-3-3} = \frac{-4}{-6} = \frac{2}{3}$$

Since $-\frac{3}{2}$ and $\frac{2}{3}$ are opposite reciprocals of one another we know that each of the angles of the figure is a right angle, so the polygon is either a rectangle or a square. Now we'll use the distance formula to figure out which of these is true.

Remember that the distance formula is $d = \sqrt{(x_1 - x_2)^2 + (y_1 - y_2)^2}$

$$d_{AB} = \sqrt{(3-7)^2 + (7-1)^2} = \sqrt{(-4)^2 + (6)^2} = \sqrt{16+36} = \sqrt{52}$$

$$d_{BC} = \sqrt{(7-1)^2 + (1--3)^2} = \sqrt{(6)^2 + (4)^2} = \sqrt{16+36} = \sqrt{52}$$

$$d_{CD} = \sqrt{(1--3)^2 + (-3-3)^2} = \sqrt{(4)^2 + (-6)^2} = \sqrt{16+36} = \sqrt{52}$$

$$d_{AD} = \sqrt{(3--3)^2 + (7-3)^2} = \sqrt{(6)^2 + (4)^2} = \sqrt{36+16} = \sqrt{52}$$

Since all four sides are the same length and all four sides meet at right angles, we have determined that this figure is a square.

Geometric Probability

Geometric probability is probability that pertains to geometric figures. Consider this: if J is the region that contains region M, so that J is inside M (like a bull's-eye). If a point K is chosen at random, then the probability that it is in region M is as follows:

P(Point K is in region M) = Area of M/Area of J.

Exercise 1

Find the probability that a point chosen at random in the figure below will lie outside the square but inside the circle.

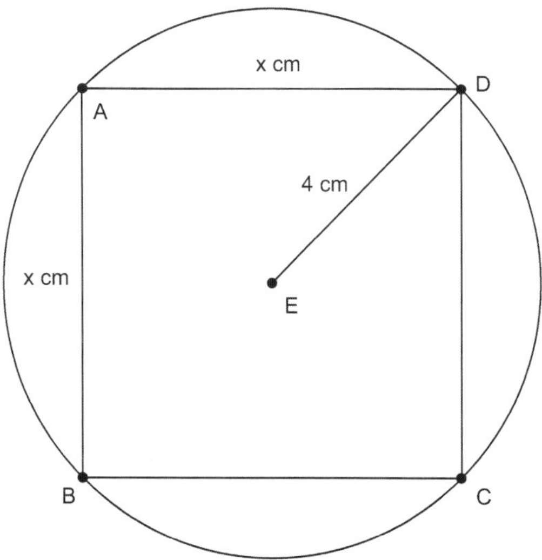

To complete this problem, you must recall the definition of probability. Put simply, the probability that something will occur is found by dividing the number of times it can occur (the area of the circle minus the area of the square in this case) by the total number of possibilities that can occur (the area of the circle in this case). Let's look at that a little more specifically in this example.

We're asked to find the probability that a point chosen at random will lie outside the square but inside the circle. That means we're looking for the total area that is included between the blue lines of the square and the green lines of the circle. To find this probability we want the number of successes (area of the circle minus area of the square which will give you the area of the spaces between the circle and the square) divided by the total possible outcomes (which is the same thing as saying anything inside the circle so the area of the circle).

In a formula, that would look like this:

$$P = \frac{\text{Area of Circle - Area of Square}}{\text{Area of Circle}}$$

So, we just need to find the area of each figure. The area of the circle is the easiest. Since we know the formula for the area of a circle is $A = \pi r^2$ all we need to do is plug in the value of our radius, which is 4.This gives us $A = \pi(4)^2 = 16\pi \approx 50.27$. Note that this calculation assumes that we used the π key on our calculator and not 3.14 as an approximation for π.

Now we need to find the area of the square. Remember that the area of a square is found by squaring the value of a side, which in this case is represented by *x*. We can use the fact that the circle has a radius of 4 to find *x*. Since the radius is equal to 4 and that value is half the distance of the diagonal of the square, the whole diagonal would be 8. The diagonal forms a right triangle with the two sides. We can use the Pythagorean Theorem to find the value of *x*. This gives us:

$x^2 + x^2 = 8^2$ ← Apply Pythagorean Theorem.

$2x^2 = 64$ ← Simplify and combine like terms.

$\frac{2x^2}{2} = \frac{64}{2}$ ← Divide both sides by 2.

$x^2 = 32$

$x = \sqrt{32} \approx 5.66$ ←To find *x* when you have x^2 you take the square root of both sides.

We're not quite done yet — remember this only gave us the value of *x*. We still need to use this to find the area of the square, which is equal to x^2. If you look back up a step you'll see that this is equal to 32.

Now that we know the area of the square and the area of the circle, we substitute both of these values back into the formula for probability we came up with above.

$$P = \frac{\text{Area of Circle - Area of Square}}{\text{Area of Circle}} = \frac{50.27 - 32}{50.27} = \frac{18.27}{50.27} \approx 0.36$$

So, the probability that a point chosen at random would lie outside the square but inside the circle is 0.36 (which is the same as 36 out of 100 or 36%).

Area, Volume, and Surface Area

The **area** of a figure is compared to covering a two-dimensional space. **Surface area** is the covering of a three-dimensional space. **Volume** is the quantity it would take to fill an object. These calculations are used in real-world applications. For instance, see the exercise below.

Exercise 1

Find the number of square meters of aluminum used to make 20,000 soda cans if the cans are the size shown below.

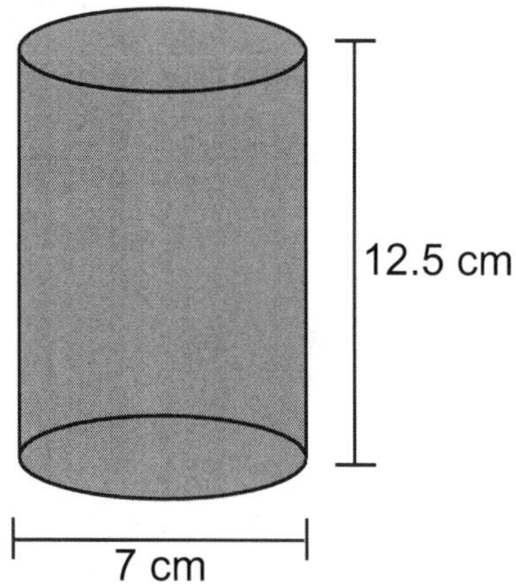

We need to find the total surface area of this shape to solve this problem. We are making some assumptions that the can is completely smooth and that we are not to account for the tab on the top of the soda can or the any other surface irregularities.

There is a formula for the surface area of a cylinder. That formula is $S.A. = 2\pi rh + 2\pi r^2$. Let's look at this formula for a moment before we work out the problem. The second part of the formula, $2\pi r^2$, is pretty easy to understand if you remember that the area of a circle is equal to πr^2. Since a cylinder like this one has two circles on it (one on top and one on bottom) if you find the area and multiply by 2 then you've found the surface area of the top and bottom.

That leaves the column or the body of the can for the other part of our formula which is $2\pi rh$. The height of our can is *h*, but what does the $2\pi r$ represent? Remember that $2\pi r$ is the formula for the distance around a circle (called the circumference). If you were to take this can and cut it down the side and lay it open, it would form a rectangle that would be *h* cm tall (in this case 12.5 cm) and would be the same width as the circumference of the circle. Since it forms a rectangle when we cut it open, to find its surface area we just multiply the circumference by the height.

Now, we just have to substitute the values into the problem. We need the radius, *r*, and the height, *h*. We have the diameter (7 cm) and *h* (12.5 cm). To find *r* we just have to divide the diameter by 2 (so 7/2), which gives us 3.5 cm. Now we can substitute these numbers into our formula and simplify:

$$S.A. = 2\pi rh + 2\pi r^2$$

$$S.A. = 2\pi(3.5)(12.5) + 2\pi(3.5)^2$$

$$S.A. = 2\pi(43.75) + 2\pi(12.25)$$

$$S.A. = 87.5\pi + 24.5\pi$$

$$S.A. = 112\pi$$

$$S.A. \approx 351.86$$

This is the surface area of a single can. Note that, once again, we used the π key on our calculator and not 3.14 as an approximation for π.

The problem asks us to find the number of square meters required to make 20,000 cans, and we have found the surface area of a single can. So, we need to multiply the surface area of 1 can times 20,000 to find the number of square centimeters required for 20,000 cans and then convert that number to square meters.

S.A. of 20,000 cans $\approx 351.86 \cdot 20000 \approx 7{,}037{,}200$ cm^2

To convert from cm^2 to m^2 you need to remember that there are 100 centimeters in a meter there are 100 x 100 cm in a square meter or 10,000 cm^2 = 1 m^2. So, if we divide 7,037,200 by 10,000 we get 703.72 m^2.

Geometric Transformations

Transformations are movements of geometric figures in the coordinate grid. The possible transformations are reflection, rotations, translation, and dilation. Consider a reflection as a flip through a mirror, a rotation as a turn about a specified point, and a translation a slide up, or right (positive) and left, or down (negative). A dilation is an enlargement or reduction by a specified scale factor. Think of enlarging or reducing digital pictures.

Rigid transformations are those that do not alter the size or the shape of the object. Reflections, rotations, and translations are all rigid transformations. A dilation alters the size of the object, so it is not a rigid transformation.

Exercise 1 (Reflection)

Reflect triangle *ABC* shown below over the *x*-axis.

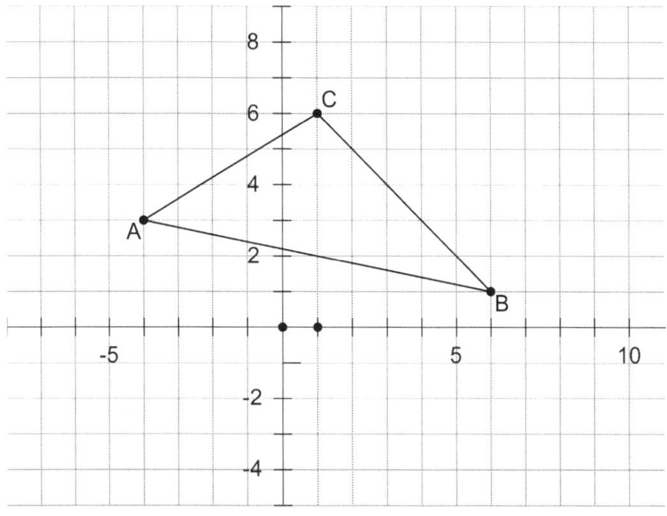

To reflect something means to draw its mirror image using the line specified. In this case, we're using the *x*-axis to reflect the figure. This means we want to plot point A 3 units below the *x* axis (since it is currently 3 units above the *x*-axis). We'll still place it at *x* = 4, but the *y* value will be -3. For point B, we'll use the same *x* value of 6, but it will now have a *y* value of -1. Finally, the reflection of point C will still have an *x* value of 1, but the *y* value will be -6. Let's take a look at these two figures plotted on the same graph.

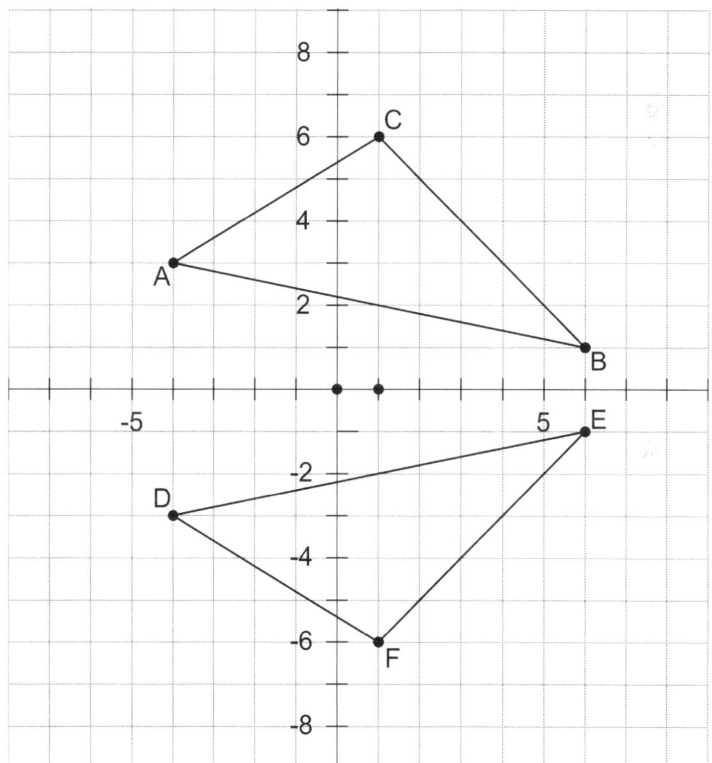

Terms You Should Know

Alternate Interior Angles - Two angles that are formed by two lines and a transversal and that lie between the two lines on opposite sides of the transversal.

Coordinate Proof - A proof that involves calculations and reference to the coordinate plane.

Consecutive Interior Angles - Two angles that are formed by two lines and a transversal that lie between the two lines on the same side of the transversal.

Distance Formula - The distance formula to find the distance between two points. The distance formula is $d = \sqrt{(x_1 - x_2)^2 + (y_1 - y_2)^2}$.

Geometric Probability - The probability that pertains to geometric figures.

Hypotenuse - The hypotenuse is always the side of a triangle located across from the right angle.

Inscribed Polygon - All vertices of the polygon touch a circle that surrounds it.

Polygon - A figure that is formed by three or more segments called sides.

Postulate - Geometric rules that are accepted without proof.

Pythagorean Theorem - The Pythagorean Theorem says that the sum of the squares of the two shorter sides of a triangle (called legs) is equal to the square of the longest side (called the hypotenuse).

Rigid Transformations - A transformation that does not alter the size or shape of a figure; rotations, reflections, translations are all rigid transformations.

Scale Factor - The ratio of the lengths of two corresponding sides of two similar polygons.

Similar Polygons - Two polygons whose corresponding angles are congruent and the lengths of the corresponding sides are proportional.

Supplementary - Sums to be 180°.

Theorem - A true statement that follows as a result of other true statements.

Transformations - The operation that maps or moves a pre-image onto an image.

Vertical Angles - Two angles whose sides form two pairs of opposite rays.

Trigonometry

The word "trigonometry" comes from the Greek roots trigon, which means "triangle," and metron, which means "measure." Simply put, trigonometry is triangle measurement—specifically the sides and the angles of triangles. Trigonometry is one of the most used branches of mathematics, as it has direct applications in navigation and electronics. However, trigonometry is not taught just using triangles anymore; modern trigonometry classes also use the circle to teach trigonometric concepts.

Trigonometry ties together many of the procedures and concepts you have learned in algebra and geometry courses. Trigonometry often is taught as an introduction to a pre-calculus course or as a standalone course for a semester with pre-calculus taught in the other semester.

Trigonometry involves not only finding measures of side lengths and angles in triangles, but also graphing trigonometric functions and working with trigonometric identities, which can be thought of as a form of algebraic proof.

Example 1: Convert Between Radians and Degrees

Convert to 30° radians. Convert $\frac{3\pi}{4}$ radians to degrees.

Before we can solve this problem, we need to understand what a degree is and what a radian is. A **degree** is equal to $\frac{1}{360}$ of a circle. Hence, this is why we say there are 360 degrees in a circle. A **radian** is the length of an arc of the circle, divided by the radius of the circle. We use a standard **unit circle** (a circle with a radius of 1) to define a radian. If you recall, the circumference of a circle is $C = 2\pi r$. So for a circle with a radius of 1 (a unit circle) the circumference is 2π . Since the circumference is the distance all the way around the circle and the whole circle is equal to 360 degrees, then we can say that 2π radians = 360 degrees or π radians = 180 degrees (divide both sides by 2).

To convert between radians and degrees, we can use the following:

To convert radians → Degrees multiply radians by $\frac{180}{\pi}$.

To convert degrees → Radians multiply degrees by $\frac{\pi}{180}$.

We can use these to complete our problem.

To convert 30° to radians we'll multiply $30 \cdot \frac{\pi}{180} = \frac{\pi}{6}$ radians.

To convert $\frac{3\pi}{4}$ radians to degrees, multiply

$\frac{3\pi}{4} \cdot \frac{180}{\pi} = \frac{3 \cdot 180}{4} = \frac{540}{4} = 135^\circ$.

Example 2: Using Triangles to Find Trig Functions

Find the sine, cosine, tangent, cosecant, secant, and cotangent of 45° and of 60° using triangles.

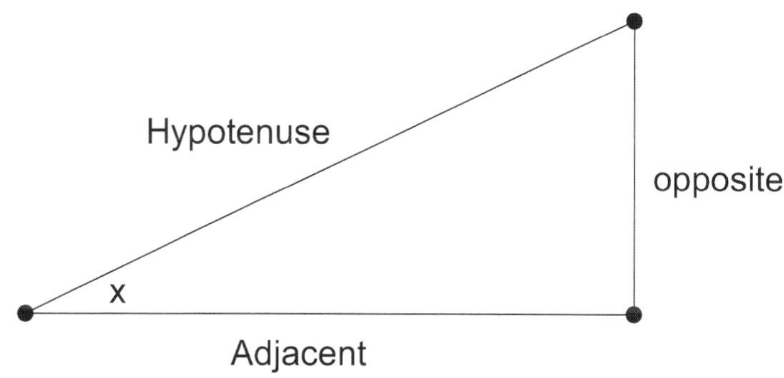

Before we dive into this problem, let's take a look at what these trigonometric functions mean. In a right triangle like the one below, we could use the angle *x* to find each of the six trigonometric functions in terms of the hypotenuse, opposite, and adjacent sides.

$$\sin x = \frac{\text{opposite}}{\text{hypotenuse}}$$

$$\cos x = \frac{\text{adjacent}}{\text{hypotenuse}}$$

$$\tan x = \frac{\text{opposite}}{\text{adjacent}}$$

Sine (sin), Cosine (cos), and Tangent (tan) are the three primary trigonometric functions. We find the other three trigonometric functions – Cosecant (csc), Secant (sec), and Cotangent (cot) – by taking the reciprocal of each of the primary trigonometric functions respectively. So, this gives us:

$$\csc x = \frac{\text{hypotenuse}}{\text{opposite}} \qquad \sec x = \frac{\text{hypotenuse}}{\text{adjacent}} \qquad \cot x = \frac{\text{adjacent}}{\text{opposite}}$$

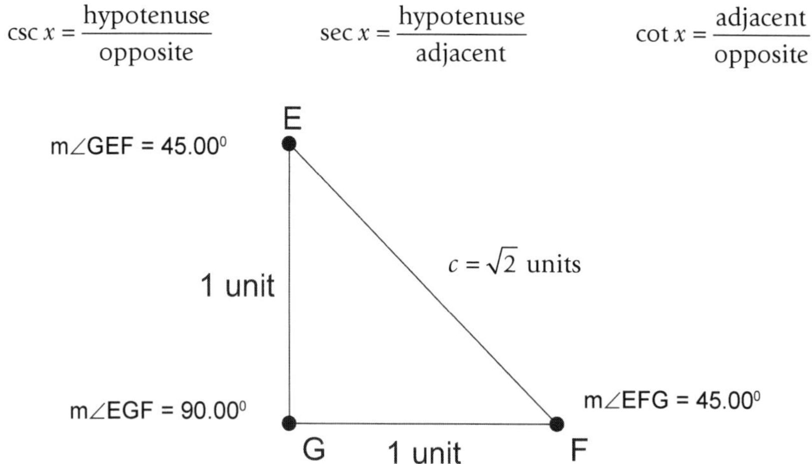

We use two standard triangles to define the most commonly used angle measures. The first is an isosceles right triangle (two sides are equal) and the second is an equilateral triangle (all three sides are equal). Let's take a look at how these help us to define trig functions for standard angles.

This is called a $45° - 45° - 90°$ right triangle. It is an isosceles triangle, which means sides EG and GF are the same length. Let each of those equal 1 unit. If we do that, then we can find the length of the hypotenuse using the Pythagorean Theorem. This gives us

$1^2+1^2=c^2$. If we solve this for c we get:

$$1^2+1^2=c^2$$

$$1+1=c^2$$

$$2=c^2$$

$$\sqrt{2}=c$$

We can use this triangle to define all six trig functions for an angle that measures 45° or $\frac{\pi}{4}$ radians. This would give us the following:

$$\sin 45^\circ=\frac{\text{opposite}}{\text{hypotenuse}}=\frac{1}{\sqrt{2}}=\frac{1\cdot\sqrt{2}}{\sqrt{2}\cdot\sqrt{2}}=\frac{\sqrt{2}}{2} \qquad \csc 45^\circ=\frac{\text{hypotenuse}}{\text{opposite}}=\sqrt{2}$$

$$\cos 45^\circ=\frac{\text{adjacent}}{\text{hypotenuse}}=\frac{1}{\sqrt{2}}=\frac{1\cdot\sqrt{2}}{\sqrt{2}\cdot\sqrt{2}}=\frac{\sqrt{2}}{2} \qquad \sec 45^\circ=\frac{\text{hypotenuse}}{\text{adjacent}}=\sqrt{2}$$

$$\tan 45^\circ=\frac{\text{opposite}}{\text{adjacent}}=\frac{1}{1}=1 \qquad \cot 45^\circ=\frac{\text{adjacent}}{\text{opposite}}=1$$

We can use another special triangle—an equilateral triangle—to find the values of trig functions for a 60° angle. Note: this triangle also can be used to find the trig functions for a 30° angle. **Think:** A $60^\circ-60^\circ-60^\circ$ triangle can be cut in half with an altitude. This will create two $30^\circ-60^\circ-90^\circ$ triangles.

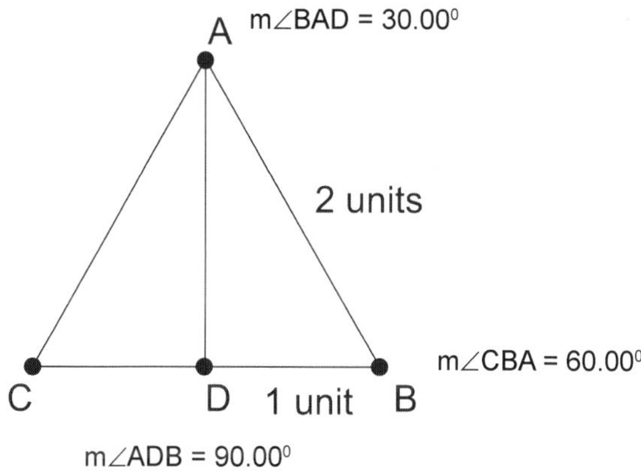

Let's draw an equilateral triangle and let each of the sides be 2. If you bisect one of the angles, it will connect to the midpoint of the other side, which will divide that side into two equal segments, with each measuring 1 unit. You can use the Pythagorean Theorem again to find the length of the segment drawn from the angle to the opposite side (AD in the picture).

$AD = \sqrt{3}$ units

$AD^2 + 1^2 = 2^2$

$AD^2 + 1 = 4$

$AD^2 = 3$

$AD = \sqrt{3}$ units

So, the ratio that corresponds to each special right triangle is:

$45° - 45° - 90°$ corresponds to $1 - 1 - \sqrt{2}$

$30° - 60° - 90°$ corresponds to $1 - \sqrt{3} - 2$

Now that we know the length of the three sides of our $30° - 60° - 90°$ triangle, we can use this to find the values of the trigonometric functions for an angle that measures $60°$. **Note:** you may need to rationalize the denominators as we did above to get the answers shown here.

$\sin 60° = \dfrac{\text{opposite}}{\text{hypotenuse}} = \dfrac{\sqrt{3}}{2}$

$\cos 60° = \dfrac{\text{adjacent}}{\text{hypotenuse}} = \dfrac{1}{2}$

$\tan 60° = \dfrac{\text{opposite}}{\text{adjacent}} = \sqrt{3}$

$\csc 60° = \dfrac{\text{hypotenuse}}{\text{opposite}} = \dfrac{2\sqrt{3}}{3}$

$\sec 60° = \dfrac{\text{hypotenuse}}{\text{adjacent}} = 2$

$\cot 60° = \dfrac{\text{adjacent}}{\text{opposite}} = \dfrac{\sqrt{3}}{3}$

You will find that as you use the trig functions on a more regular basis, you will get to know some of the values for trig functions of certain angles. It's also a good idea to study the method we used to create the two right triangles (the 30-60-90 and 45-45-90) so you can redo that yourself if you need to recall the values of the trig functions. A table like this one also may be helpful to keep handy.

Degrees	Radians	sin θ	cos θ	tan θ	csc θ	sec θ	cot θ
0	0	0	1	0	Undefined	1	Undefined
30	$\pi/6$	1/2	$\sqrt{3}/2$	$\sqrt{3}/3$	2	$2\sqrt{3}/3$	$\sqrt{3}$
45	$\pi/4$	$\sqrt{2}/2$	$\sqrt{2}/2$	1	$\sqrt{2}$	$\sqrt{2}$	1
60	$\pi/3$	$\sqrt{3}/2$	1/2	$\sqrt{3}$	$2\sqrt{3}/3$	2	$\sqrt{3}/3$
90	$\pi/2$	1	0	Undefined	1	Undefined	0
120	$2\pi/3$	$\sqrt{3}/2$	-1/2	$-\sqrt{3}$	$2\sqrt{3}/3$	-2	$-\sqrt{3}/3$
135	$3\pi/4$	$\sqrt{2}/2$	$-\sqrt{2}/2$	-1	$\sqrt{2}$	$-\sqrt{2}$	-1
150	$5\pi/6$	1/2	$-\sqrt{3}/2$	$-\sqrt{3}/3$	2	$-2\sqrt{3}/3$	$-\sqrt{3}$
180	π	0	-1	0	Undefined	-1	Undefined
210	$7\pi/6$	-1/2	$-\sqrt{3}/2$	$\sqrt{3}/3$	-2	$-2\sqrt{3}/3$	$\sqrt{3}$
225	$5\pi/4$	$-\sqrt{2}/2$	$-\sqrt{2}/2$	1	$-\sqrt{2}$	$-\sqrt{2}$	1
240	$4\pi/3$	$-\sqrt{3}/2$	-1/2	$\sqrt{3}$	$-2\sqrt{3}/3$	-2	$\sqrt{3}/3$
270	$3\pi/2$	-1	0	Undefined	-1	Undefined	0
300	$5\pi/3$	$-\sqrt{3}/2$	1/2	$-\sqrt{3}$	$-2\sqrt{3}/3$	2	$-\sqrt{3}$
315	$7\pi/4$	$-\sqrt{2}/2$	$\sqrt{2}/2$	-1	$-\sqrt{2}$	$\sqrt{2}$	-1
330	$11\pi/6$	-1/2	$\sqrt{3}/2$	$-\sqrt{3}/3$	-2	$2\sqrt{3}/3$	$-\sqrt{3}$
360	2π	0	1	0	Undefined	1	Undefined

Example 3: Using the Unit Circle to Find the Trig Functions

Use the unit circle to find the values of six trig functions for $5\pi/6$ radians.

We could simply look in the table above and read off the values for the six trig functions for $5\pi/6$ radians. However, to gain a deeper understanding of the six trig functions, let's take a look at the unit circle and see how it can be used to find the values of trig functions.

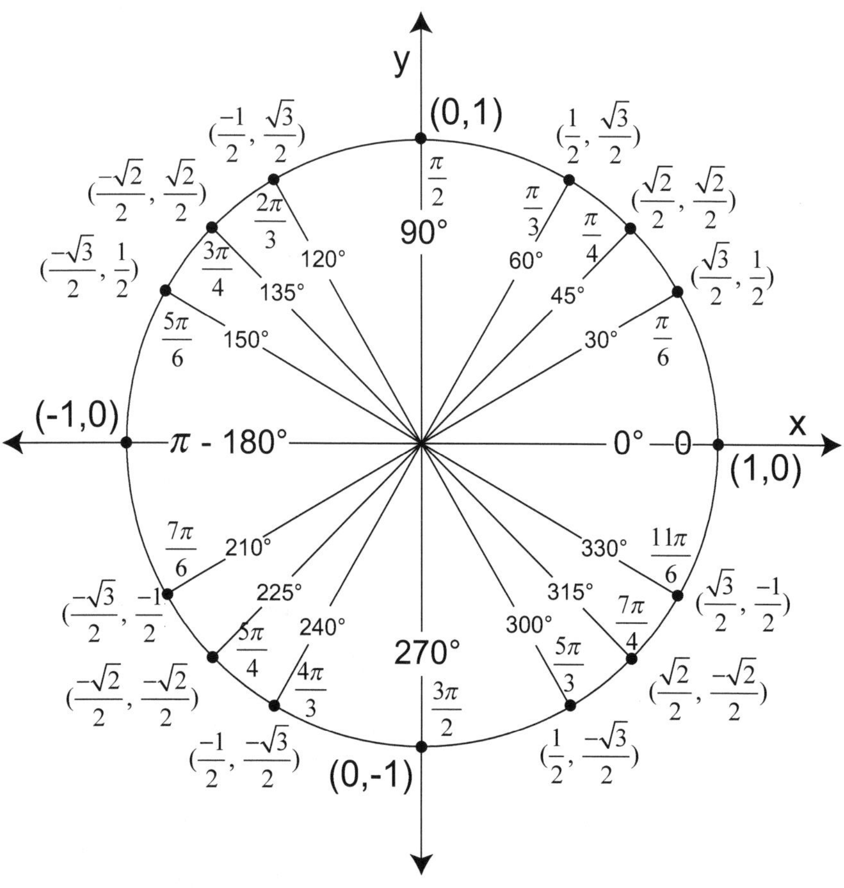

The Unit Circle is a circle with radius of 1 unit, and centered at the origin. If you recall, this would give the circle an equation of $x^2 + y^2 = 1$. Remember also that a radian is defined as the length of the arc divided by the radius. In the unit circle, you will always divide by 1, so we can define each of the radian values by just their arc length. On the unit circle, we always measure counterclockwise from the *x*-axis. We can determine from this that the values of the trig functions, when you're using the unit circle, can be defined as follows:

$$\sin\theta = \frac{y}{r} \qquad \csc\theta = \frac{r}{y}$$

$$\cos\theta = \frac{x}{r} \qquad \sec\theta = \frac{r}{x}$$

$$\tan\theta = \frac{y}{x} \qquad \cot\theta = \frac{x}{y}$$

Remember *r* is always 1 and the values of *x* and *y* can be found by looking at the coordinates of the point that correspond to each angle on the circle. To find the value of the trig functions for $5\pi/6$, we would find $5\pi/6$ on the unit circle and the substitute the values of the point $\left(-\sqrt{3}/2, 1/2\right)$ and the value of *r*, which is 1, into the formulas for each of the trig functions. This would give us:

$$\sin\theta = \frac{1}{2}$$

$$\cos\theta = -\frac{\sqrt{3}}{2}$$

$$\tan\theta = \frac{\frac{1}{2}}{-\frac{\sqrt{3}}{2}} = \frac{1}{2}\cdot -\frac{2}{\sqrt{3}} = -\frac{2}{2\sqrt{3}} = -\frac{2}{2\sqrt{3}}\cdot\frac{\sqrt{3}}{\sqrt{3}} = -\frac{2\sqrt{3}}{2\cdot 3} = -\frac{\sqrt{3}}{3}$$

$$\csc\theta = 2$$

$$\sec\theta = \frac{1}{-\frac{\sqrt{3}}{2}} = 1\cdot -\frac{2}{\sqrt{3}} = -\frac{2}{\sqrt{3}} = -\frac{2}{\sqrt{3}}\cdot\frac{\sqrt{3}}{\sqrt{3}} = -\frac{2\sqrt{3}}{3}$$

$$\cot\theta = \frac{-\frac{\sqrt{3}}{2}}{\frac{1}{2}} = -\frac{\sqrt{3}}{2}\cdot\frac{2}{1} = -\sqrt{3}$$

THINK: To create the unit circle, you would need to use some common math sense and some trigonometry. For instance, quadrants II and III on the coordinate grid are negative for *x* and quadrants III and IV are negative for *y*. Also, consider the *x*-axis and the *y*-axis lines of symmetry with the pre-image in quadrant I. Notice the first quadrant and its radians are reflected throughout the unit circle in each quadrant. The coordinates of quadrant I are repeated in each quadrant with changes in the + and – values dictated by the quadrant.

Example 4: Trigonometric Identities

Show that $(1-\tan x)^2 = \sec^2 x - 2\tan x$.

A **trigonometric identity** is a form of proof in which you use known properties of the trig functions as well as known identities of the trig functions to show that other trig identities are true. You already know the relationships between sin and csc, cos and sec, and tan and cot, but we'll write them again here for reference.

$$\sin x = \frac{1}{\csc x} \qquad \csc x = \frac{1}{\sin x}$$

$$\cos x = \frac{1}{\sec x} \qquad \sec x = \frac{1}{\cos x}$$

$$\tan x = \frac{1}{\cot x} \qquad \cot x = \frac{1}{\tan x}$$

In addition, tan x = $\frac{\sin x}{\cos x}$ and cot x = $\frac{\cos x}{\sin x}$

Since we're dealing with right triangles in our study of trigonometry, you might also notice there is something called a Pythagorean Identity in trigonometry. The Pythagorean Identity comes in handy; it says $\sin^2 x + \cos^2 x = 1$. From this identity, you can either divide each term by $\sin^2 x$ or by $\cos^2 x$ to come up with these other two identities:

$$\tan^2 x + 1 = \sec^2 x \qquad \text{and} \qquad \cot^2 x + 1 = \csc^2 x$$

So, our three Pythagorean Identities are:

$$\sin^2 x + \cos^2 x = 1 \qquad \tan^2 x + 1 = \sec^2 x \qquad \cot^2 x + 1 = \csc^2 x$$

The challenge with trigonometric identities is knowing which side to work with and which side to leave alone. In many cases, you'll need to start over a couple of times before you find the path that leads you to the solution. A few rules of thumb that often work (but in this case don't necessarily get you to the solution quickly) are to try the following:

1. Work with the side that appears to be the "messiest" (i.e. the side with the most terms or the side with the squared terms). It is often easier to simplify a "messy" side of the equation to make it equal to the "simpler" side.

2. Change tan to sin and cos and cot to sin and cos.

3. Sometimes it is useful to write everything in terms of sin and cos (eliminate tan, cot, sec, csc).

These are not absolute rules, but they tend to work more often than not. However, in the case of our example they actually don't help us much at all. Let's take a look at our example: $(1-\tan x)^2 = \sec^2 x - 2\tan x$

If we were to apply the guidelines above we would probably work with the right-hand side of the equation first, as it appears a little more complex than the left and we would change sec and tan to terms of sin and cos. While that may work, it is not the most efficient way to get to the solution. Let's start with the left side (we're going to try to turn the left side into the right side to show that this is a true statement).

$(1-\tan x)^2 = (1-\tan x)(1-\tan x)$ ←Write out the two factors.

$= 1 - \tan x - \tan x + \tan^2 x$ ←Multiply the factors using FOIL.

$= 1 - 2\tan x + \tan^2 x$ ←Combine like terms.

Let's pause for a moment and take a look at where we are. We have a -2 tan *x*, which we want to have in our answer. We know from the Pythagorean Identity above that $\tan^2 x + 1 = \sec^2 x$ so we can rearrange the terms just a bit and make a substitution and we have what we wanted. Let's do that now.

$= 1 + \tan^2 x - 2\tan x$ ←Rewrite terms using the commutative property.

$= \sec^2 x - 2\tan x$ ←Substitute using a Pythagorean Identity.

Example 5: Graphing Trig Functions

Graph $y = 3 + \sin x$

Graphing a trig function can be easily done with a graphing calculator, but you can also graph by hand. Typically, you let radian values run along the x axis of the graph and unit values run along the y axis of the graph.

You can graph the trig function by making a chart of values, substituting in standard radian measures like $\frac{\pi}{4}$, $\frac{\pi}{2}$, π, etc. in for x, and finding out the value of y. When you do this, you get the graph shown below.

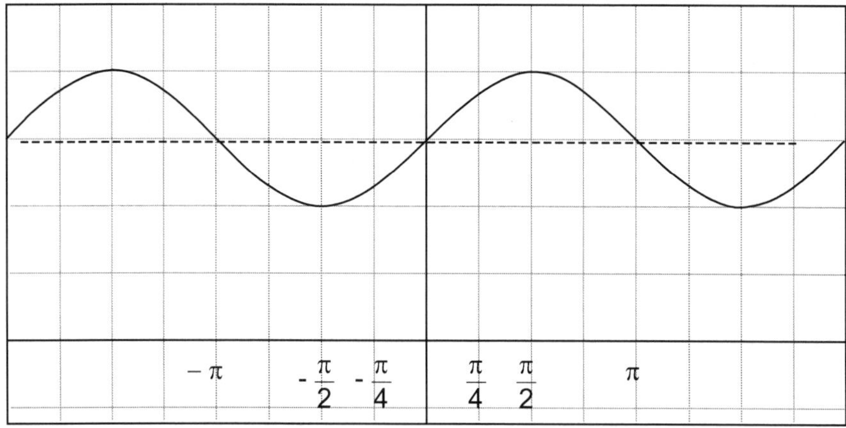

Notice that this graph appears to repeat itself — making one full cycle in the span of 2π. This is called the graph's **period,** or the length it takes for the graph to run a full cycle. The period runs from $[0, 2\pi]$. Also notice the dashed green line on the graph. Note that the graph is moving about that line and that this line is located three units above the x-axis. Recall that our equation was $y = 3 + \sin x$. The 3 added to the front moved the graph of $\sin x$ up three units. The graph moves one full unit above that line and one full unit below that line so it is said to have an **amplitude** of 1 (the highest peak it reaches above or below the line).

Exercise 1

Find the value of $\cot 180^\circ$.

You can use a calculator, the table above, or the unit circle to find this answer. Remember that cot x= $\frac{1}{\tan x}$ or $\frac{\cos x}{\sin x}$. If we look at 180° on the unit circle, x = -1 and y = 0. This gives us cot 180 = $\frac{-1}{0}$, since the cot = x/y. We know that any time you divide by 0 the answer is undefined, therefore cot 180 is undefined.

Exercise 2

Find the angle x that has $\sin x = \frac{-\sqrt{2}}{2}$.

Use a right triangle and the concept of Soh Cah Toa to help you label the opposite side as $-\sqrt{2}$ and the hypotenuse as 2.

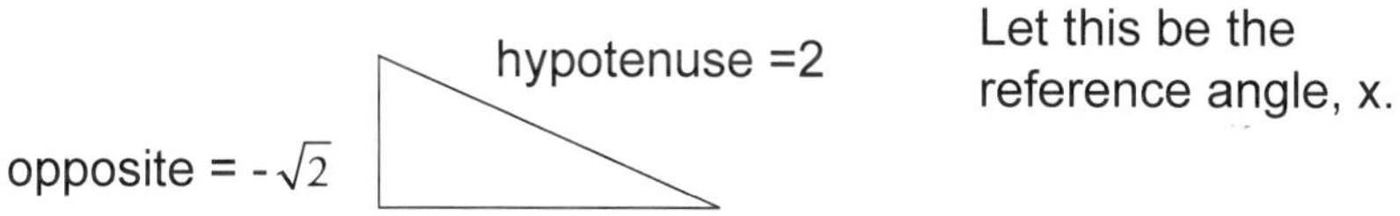

hypotenuse =2

Let this be the reference angle, x.

opposite = - $\sqrt{2}$

This problem is easiest to do from the table of values that we created above. From that table it is easy to see that both $\frac{5\pi}{4}$ and $\frac{7\pi}{4}$ have a $\sin x = \frac{-\sqrt{2}}{2}$. Finding the angle given the sin value is called the arc sin function.

Before we move on to the next exercise, there is another important fact taught in most trigonometry classes: The sine function is positive in the first and second quadrants of the graph, so we know that this must lie in either the third or the fourth quadrant for it to be true (which it does). Each of the trig functions is positive in exactly two quadrants. You can remember which quadrants using the sentence “All Students Take Calculus.” Notice that each word of this little sentence starts with A, S, T, and C, which stand for All positive (in the first quadrant), Sin positive (in the second quadrant), Tan positive (in the third quadrant), and Cos positive (in the fourth quadrant). You can also remember it by looking at a graph and writing in the letters or the names of the functions like this:

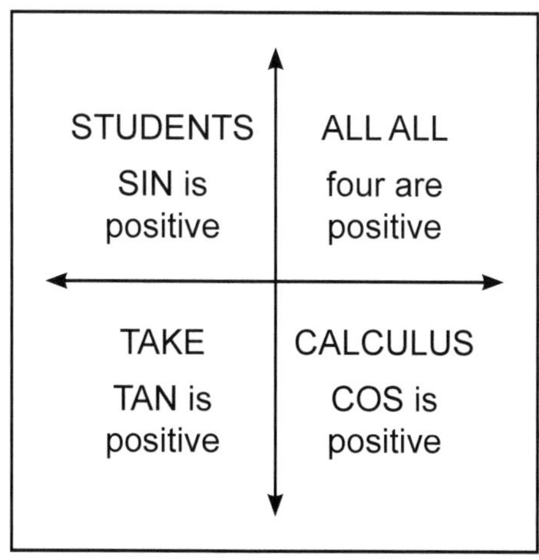

Exercise 1

Find $\tan^{-1}(-0.45)$.

There are some problems that are best worked on a calculator, and this is one of them. If you don't have a calculator available, use a table of values in the back of trigonometry textbooks (just look for $\tan^{-1}$ or arctan tables in the index or Table of Contents). However, the easiest way to do this problem is to enter it into your calculator, which will give a result of -0.423.

Exercise 2

Find $\sin^{-1}\left(\frac{1}{\sqrt{2}}\right)$.

Remember that the inverse of a trig function is asking for the angle at which that trig function occurs. There are a variety of ways we can find this answer, but the easiest ways are to look back at the table of values, use the unit circle, or use the calculator. Recall that $\frac{1}{\sqrt{2}} = \frac{\sqrt{2}}{2}$ and $\frac{\pi}{4}$ is the angle associated with this value. So the answer is $\frac{\pi}{4}$.

Exercise 3

Solve $\cos x = \frac{1}{\sqrt{2}}$.

Unlike the inverse function above, where the inverse only exists in certain quadrants, this is an equation that must be solved. Therefore, since our answer is positive and we know that cosine is positive in the first and fourth quadrants we may have an answer in each of these quadrants. Again, this is easiest to do by looking back at the unit circle or the table of values and finding where cosine is equal to $\frac{1}{\sqrt{2}}$. Remember that this may be written as either $\frac{1}{\sqrt{2}}$ or as $\frac{\sqrt{2}}{2}$. We see that the *x* value is equal to $\frac{\sqrt{2}}{2}$ in the first quadrant at $\frac{\pi}{4}$ and in the fourth quadrant at $\frac{7\pi}{4}$, so these are the two possible values of *x*.

Exercise 4

Solve $\sec x = \frac{5}{4}$.

This one is a bit different because there are no values in our tables that look like $\frac{5}{4}$, so we'll use some of the same thinking but a slightly different method here. Let's start by rewriting this as $\cos x = \frac{4}{5}$, because cosine is slightly easier to work with than secant. Now, recall in which quadrants cosine is positive. Cosine is positive in the first and the fourth quadrants so there will be two answers, one for each of those two quadrants.

Also, remember that when we defined the unit circle values we said that cosine was equal to $\frac{x}{r}$. While we used that to define the unit circle, it holds true for any circle. So, if we use that to think about our problem here, we have an *x* value of 4 and a radius of 5.

Let's sketch that on a graph in both the first and fourth quadrants.

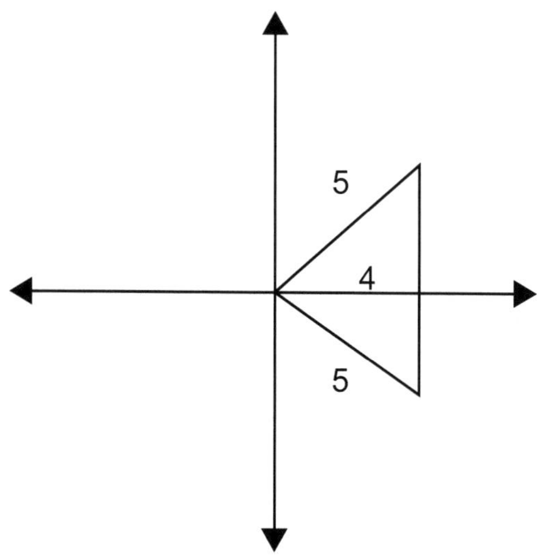

Two right triangles are formed, with the value along the *x*-axis being four units and the hypotenuse of each being 5.

We can use the inverse function on the calculator to find the value in the first quadrant, which is $\cos^{-1}\frac{4}{5} = 0.644$. Now we need to use a little logic to figure out the value in the fourth quadrant. We know that if you travel all the way around the unit circle, you've traveled 2π radians or $2 \cdot 3.14 = 6.28$ radians. We can find our value in the fourth quadrant by subtracting the value we got from 6.28. So the value in the fourth quadrant is 6.28 – 0.644 = 5.636. Therefore *x* = 0.644 or 5.636.

TRY IT!

1. Find the value of cot 90°.

2. Find the value of sin 270°

3. Find the angle x that has $\cos x = \frac{-\sqrt{2}}{2}$.

Answers

1. You can use a calculator, the table above, or the unit circle to find this answer. Remember that $\cot x = \frac{1}{\tan x}$ or $\frac{\cos x}{\sin x}$. If we look at 90° on the unit circle, $x = 0$ and $y = -1$. This gives us $\cot 90^\circ = \frac{0}{-1}$, since the $\cot = \frac{x}{y}$. So, $\cot 90^\circ = 0$.

2. You can use a calculator, the table above, or the unit circle to find this answer. Remember that $\sin\theta = \frac{y}{r}$. If we look at 270°on the unit circle, $x = 0$ and $y = -1$. This gives us $\sin\theta = \frac{-1}{1}$. So, the $\sin\theta = -1$.

3. Draw a right triangle and label the sides as a reference for this problem.

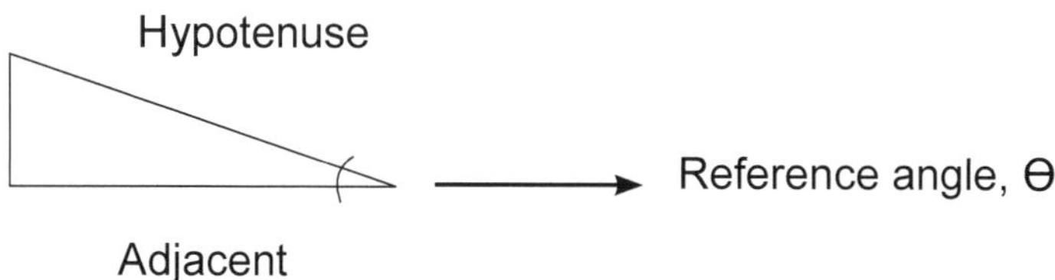

$$\cos x = \frac{-\sqrt{2}}{2} = \frac{adjacent}{hypotenuse}$$

If cos is negative, the triangle must live in quadrants II or III.

Now we will use the Pythagorean Theorem to find the missing side length, or the length of the opposite side.

$$c^2 = a^2 + b^2$$
$$2^2 = (-\sqrt{2})^2 + b^2$$
$$4 = 2 + b^2$$
$$2 = b^2$$
$$\sqrt{2} = b$$

So the triangle has two congruent sides, and thus two congruent angles. These facts make this a 45°-45°-90° triangle. So the missing angle must be a 45° angle.

Exercise 5

Find $\csc\left(-\frac{\pi}{4}\right)$.

We know from the table of values above (or the unit circle or the triangles) that the csc $\frac{\pi}{4} = \sqrt{2}$. But what happens if we want to find $\csc\left(-\frac{\pi}{4}\right)$? Once again, we have a handy set of identities that can help us. They are called the Negative Identities. They are:

$\sin(-x) = -\sin x$	$\csc(-x) = -\csc x$
$\cos(-x) = \cos x$	$\sec(-x) = \sec x$
$\tan(-x) = -\tan x$	$\cot(-x) = -\cot x$

We can see from this that we are simply going to use – csc x for our answer which is going to be $-\sqrt{2}$.

Exercise 6

Find $\sin\frac{\pi}{12}$.

$\frac{\pi}{12}$ is not one of the angles we have in our table, it's not on our unit circle, and there is no triangle relationship that can quickly give us the answer. We can always grab the calculator and get a decimal approximation of the answer, but what if we want an exact answer? Yet again, there is a set of identities that can help us, and they are called the Half-Angle Identities. We know that $\frac{\pi}{12}=\frac{1}{2}\cdot\frac{\pi}{6}$. We can use the Half-Angle Identities and this fact to find the value of $\sin\frac{\pi}{12}$. The Half-Angle Identities are:

$$\sin\frac{\theta}{2}=\pm\sqrt{\frac{1-\cos\theta}{2}}$$

$$\cos\frac{\theta}{2}=\pm\sqrt{\frac{1+\cos\theta}{2}}$$

$$\tan\frac{\theta}{2}=\frac{\sin\theta}{1+\cos\theta}\text{ OR }\frac{1-\cos\theta}{\sin\theta}$$

Notice that each half angle identity yields two answers. We must remember (using that helpful sentence "All Students Take Calculus") in which quadrants the specific trig function is positive and identify where our angle lies. Since $\frac{\pi}{12}$ lies in the first quadrant, we'll want the positive answer when we use the half-angle identity. So, to use the half-angle identities above, we'll let $\theta=\frac{\pi}{6}$. So:

$$\sin\frac{\theta}{2}=\sin\frac{\frac{\pi}{6}}{2}=\sin\frac{\pi}{12}=\pm\sqrt{\frac{1-\cos\frac{\pi}{6}}{2}}$$ ← Substitute $\theta=\frac{\pi}{6}$.

$$=\sqrt{\frac{1-\frac{\sqrt{3}}{2}}{2}}$$ ←Substitute the value of $\cos\frac{\pi}{6}$ and drop the $\pm$since we are only interested in the positive answer.

$$=\sqrt{\frac{\frac{2-\sqrt{3}}{2}}{2}}$$ ←You could stop at the step above. All remaining steps shown from here below are just simplification. In this step, we found a common denominator of 2 in the numerator and combined the two fractions. Now, we'll divide by the 2 in the denominator.

$$=\sqrt{\frac{\frac{2-\sqrt{3}}{2}}{2}}=\sqrt{\frac{2-\sqrt{3}}{2}\cdot\frac{1}{2}}=\sqrt{\frac{2-\sqrt{3}}{4}}$$

←At this point you really can't simplify any further, so this would be another way to write the answer above.

TRY IT!

1. Verify the identity $\cos\Theta(\sec\Theta-\cos\Theta)=\sin^2\Theta$.

2. Verify the identity $\frac{\cos\Theta}{1-\sin\Theta}=\sec\Theta+\tan\Theta$.

Answers

1. The left-hand side (LHS) looks more complicated, so we'll start with it and try to transform it into the right hand side (RHS).

$$LHS=\cos\Theta(\sec\Theta-\cos\Theta)$$
$$=\cos\Theta\left(\frac{1}{\cos\Theta}-\cos\Theta\right)$$
$$=1-\cos^2\Theta$$
$$=\sin^2\Theta=RHS$$
Done!

2. We start with the LHS and multiply the numerator and denominator by 1 + sinΘ.

$$
\begin{aligned}
LHS &= \frac{\cos\Theta}{1-\sin\Theta} \\
&= \frac{\cos\Theta}{1-\sin\Theta}\cdot\frac{1+\sin\Theta}{1+\sin\Theta} \\
&= \frac{\cos\Theta(1+\sin\Theta)}{1-\sin^2\Theta} \\
&= \frac{\cos\Theta(1+\sin\Theta)}{\cos^2\Theta} \\
&= \frac{1+\sin\Theta}{\cos\Theta} \\
&= \frac{1}{\cos\Theta}+\frac{\sin\Theta}{\cos\Theta} \\
&= \sec\Theta+\tan\Theta \\
&= RHS
\end{aligned}
$$

Done!

Exercise 7

Show that $\frac{1-\sin x}{\cos x}=\frac{\cos x}{1+\sin x}$.

Just like in the example we did above, the rules of thumb that are often helpful don't give us much to work with here—neither side looks more complex than the other and everything is already written in terms of sin *x* and cos *x*. In this exercise, we're going to use a little trick, which is to multiply by 1, but it is 1 in a special form. We're going to work with the left-hand side and we're going to multiply it by 1 in the form of $\frac{1+\sin x}{1+\sin x}$. **Think:** A number's **conjugate** is known as a + b = a – b, so that $(a + b)(a - b) = a^2 - b^2$. The trick of multiplying by 1 in some form is commonly used when you need to introduce a different factor into the problem and since multiplying something by 1 doesn't change it then it is perfectly acceptable to do. Let's proceed with this problem.

$\frac{1-\sin x}{\cos x}\cdot\frac{1+\sin x}{1+\sin x}$ ←Multiply by 1 in the form of $\frac{1+\sin x}{1+\sin x}$.

$=\frac{(1-\sin x)(1+\sin x)}{\cos x(1+\sin x)}$ ←Rewrite as a single fraction.

$=\dfrac{(1-\sin x)(1+\sin x)}{\cos x(1+\sin x)}$ ←Multiply the numerator using FOIL.

$=\dfrac{1+\sin x-\sin x-\sin^2 x}{\cos x(1+\sin x)}$ ←Combine like terms in the numerator.

Let's pause for a moment and recall that $\sin^2 x+\cos^2 x=1$. If we subtract $\sin^2 x$ from both sides of that equation, we get $\cos^2 x=1-\sin^2 x$ which is what we have in the numerator so we can substitute.

$=\dfrac{\cos^2 x}{\cos x(1+\sin x)}$ ←Substitute.

$=\dfrac{\cos x}{1+\sin x}$ ←Cancel $\cos^2 x$ with $\cos x$. Done!

Exercise 8

Show that $\dfrac{\tan^2 x}{\sec x+1}=\dfrac{1-\cos x}{\cos x}$.

Unlike in our previous examples, our rules of thumb are going to help us out here a bit. First of all, we're going to work with the left side, which appears slightly more complex than the right-hand side. Also, at some point we will need to change the tan and sec to terms of sin and cos, but we're not going to do that right away. Let's first recall the Pythagorean Identity that involves $\tan^2 x$, which says $\sec^2 x=1+\tan^2 x$. If we use just a little bit of algebra we can rewrite that by subtracting 1 from both sides, as $\sec^2 x-1=\tan^2 x$. Let's substitute that in for the numerator of the left side.

$\dfrac{\tan^2 x}{\sec x+1}$

$=\dfrac{\sec^2 x-1}{\sec x+1}$ ←Substitute using Pythagorean Identity.

$=\dfrac{(\sec x+1)(\sec x-1)}{\sec x+1}$ ←Factor the numerator as a difference of squares.

$= \sec x - 1$ ←Cancel sec x + 1 in the numerator and denominator.

$= \frac{1}{\cos x} - 1$ ←Rewrite sec x in terms of cos x.

$= \frac{1}{\cos x} - \frac{\cos x}{\cos x}$ ←Rewrite 1 as $\frac{\cos x}{\cos x}$.

$= \frac{1 - \cos x}{\cos x}$ ←Simplify.

Exercise 9

Graph $y = \sin\left(x - \frac{\pi}{4}\right)$.

Again, you can choose to make a table of values of x using common angles and then find y, or you can use a calculator to graph the equation. Let's make a table of values:

X	$\left(x - \frac{\pi}{4}\right)$	$\sin\left(x - \frac{\pi}{4}\right)$
-2π	$-\frac{9\pi}{4}$	-0.707
$-\frac{3\pi}{2}$	$-\frac{7\pi}{4}$	0.707
$-\pi$	$-\frac{5\pi}{4}$	0.707
$\frac{\pi}{4}$	$-\frac{3\pi}{4}$	-0.707
0	$-\frac{\pi}{4}$	-0.707
$\frac{\pi}{2}$	$\frac{\pi}{4}$	0.707

X	$\left(x-\frac{\pi}{4}\right)$	$\sin\left(x-\frac{\pi}{4}\right)$
π	$\frac{3\pi}{4}$	0.707
$\frac{3\pi}{2}$	$\frac{5\pi}{4}$	-0.707
2π	$\frac{7\pi}{4}$	-0.707

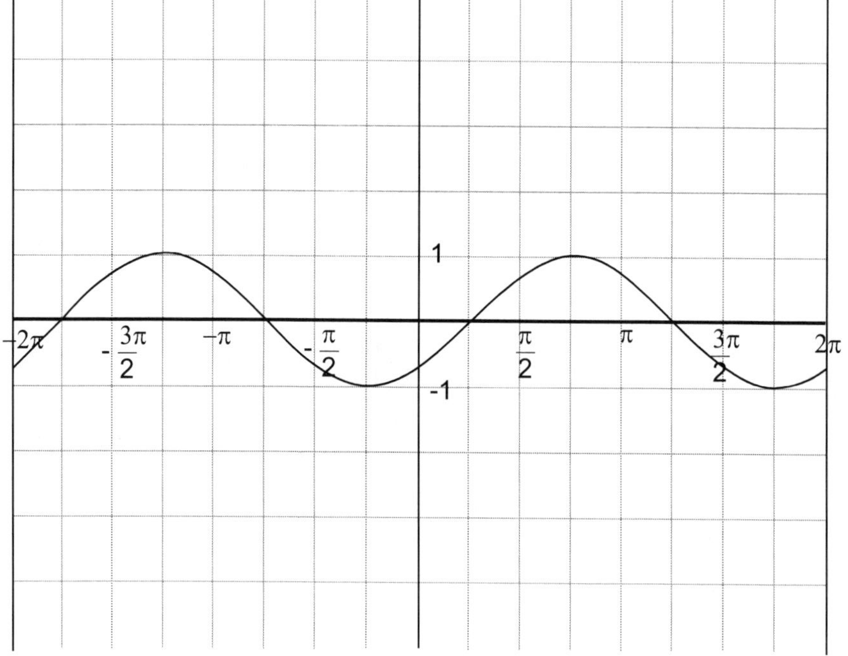

Exercise 10

Graph $y = 2\cos x$.

Again, you can use a table of values or a graphing calculator to sketch this graph. We'll make a table of values.

x	cos x	2cos x
-2π	1	2
$-\frac{3\pi}{2}$	0	0
$-\pi$	-1	-2
$-\frac{\pi}{2}$	0	0
0	1	2
$\frac{\pi}{2}$	0	0
π	-1	-2
$\frac{3\pi}{2}$	0	0
2π	1	2

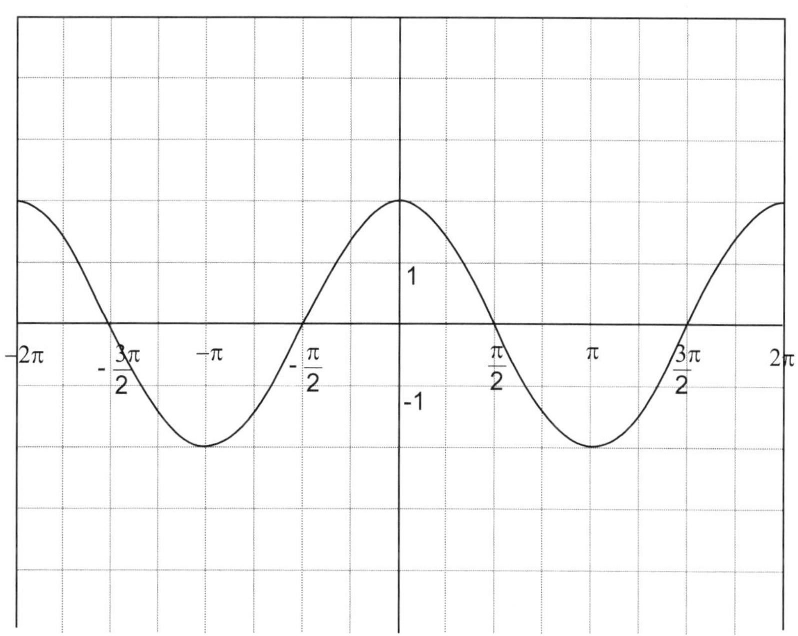

Exercise 11

Graph y = tan x.

Recall that the tan is undefined at $\frac{\pi}{2}$ and $\frac{3\pi}{2}$. (You can look back at the table of values we developed to see that this is true.) When a function is undefined at a certain value on a graph, the function is said to have an asymptote at that value. We draw asymptotes on the graph with dashed lines. We'll draw those lines when we sketch the graph. Let's fill in our table of values like we've done before.

X	tan x
-2π	0
$-\frac{3\pi}{2}$	Undefined
$-\pi$	0
$-\frac{\pi}{2}$	Undefined
0	0
$\frac{\pi}{2}$	Undefined
π	0
$\frac{3\pi}{2}$	Undefined
2π	0

This doesn't really give us enough information, so we're going to need to fill in a few more values. At $\frac{\pi}{4}$ we know tan x = 1, and at $\frac{3\pi}{2}$ we know that tan x = -1. This gives us the values for tan -$\frac{\pi}{4}$ = -1 and tan - $\frac{3\pi}{4}$ = 1. We can now sketch the graph.

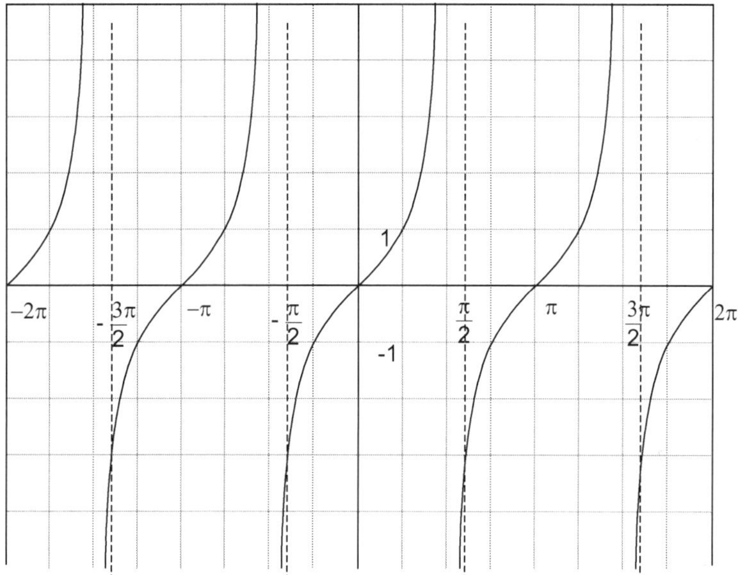

TRY IT!

1. Graph $y = \cos x$ for $-2\pi \le x \le 2\pi$

2. Graph $y = \sin x$ for $-2\pi \le x \le 2\pi$

Answers

1. Create a table of values, and then plot the points:

x	***y* = cos *x***
-2π	1
$-3\pi/2$	0
$-\pi$	-1
$-\frac{\pi}{2}$	0
0	1
$\frac{\pi}{2}$	0

x	$y = \cos x$
π	-1
$3\pi/2$	0
2π	1

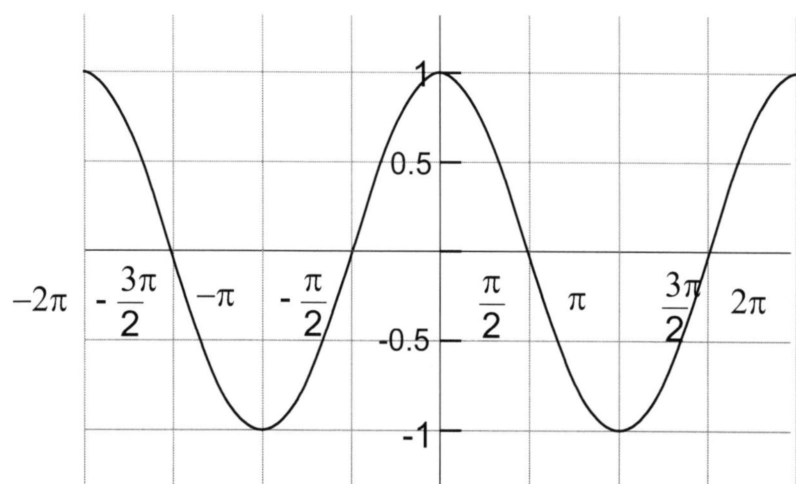

2. Create a table of values, and then plot the points:

x	$y = \sin x$
-2π	0
$-3\pi/2$	1
$-\pi$	0
$-\frac{\pi}{2}$	-1
0	0
$\frac{\pi}{2}$	1
π	0
$3\pi/2$	-1
2π	0

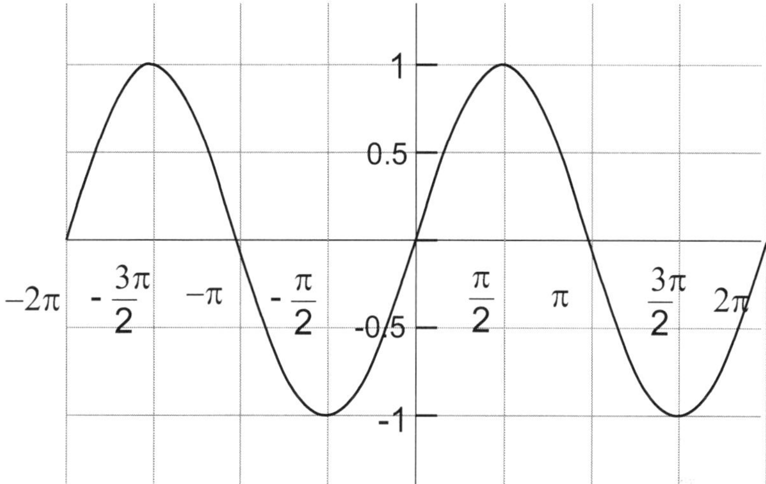

Exercise 12

Solve $\sin 2x = \cos x$.

To solve this problem, we're going to need to use another set of identities called the Double-Angle Identities. There are several different Double-Angle Identities. They are:

$$\sin 2\theta = 2\sin\theta\cos\theta$$

$$\cos 2\theta = 1 - 2\sin^2\theta$$

$$\cos 2\theta = 2\cos^2\theta - 1$$

$$\cos 2\theta = \cos^2\theta - \sin^2\theta$$

$$\tan 2\theta = \frac{2\tan\theta}{1-\tan^2\theta}$$

-2π $\quad -\frac{3\pi}{2}$ $\quad -\pi$ $\quad -\frac{\pi}{2}$ $\quad \frac{\pi}{2}$ $\quad \pi$ $\quad \frac{3\pi}{2}$ $\quad 2\pi$

We will use the Double-Angle Identity for sin 2*x* and substitute that into the equation:

$2\sin x\cos x = \cos x$

$2\sin x\cos x - \cos x = 0$ ←Substitute using the Double-Angle Identity for sin 2*x*.

$2\sin x\cos x - \cos x = 0$ ←Subtract cos *x* from both sides to get the equation equal to 0.

$\cos x(2\sin x - 1) = 0$ ←Factor out cos *x*.

Now, this is no different than solving $x(2x - 1) = 0$ back in Algebra I, where you would solve both $x = 0$ and $2x - 1 = 0$. So, we have:

$\cos x = 0$ or $2 \sin x - 1 = 0$

$\cos x = 0$ at $\frac{\pi}{2}$ and $\frac{3\pi}{2}$

$$2\sin x = 1$$

$$\sin x = \frac{1}{2}$$

$\sin x = \frac{1}{2}$ at $\frac{\pi}{6}$ and $\frac{5\pi}{6}$

So, we have four answers to the problem. They are $\frac{\pi}{2}$, $\frac{3\pi}{2}$, $\frac{\pi}{6}$, and $\frac{5\pi}{6}$.

Exercise 13

Solve $2\cos^2 x - 7\cos x + 3 = 0$

If you look closely at this problem, you might see another remnant from your Algebra I and Algebra II studies—the quadratic equation $2x^2 - 7x + 3 = 0$. If you recall from Algebra I and Algebra II, we would have factored that and then solved each of the resulting equations. That same process will work with the trig functions in place of the variable *x*.

$$2\cos^2 x - 7\cos x + 3 = 0$$

$(2\cos x - 1)(\cos x - 3) = 0$ ←Factor (remember to check using FOIL).

$2\cos x - 1 = 0$ or $\cos x - 3 = 0$

Simplify each of these to get the trig function on one side.

$2\cos x = 1$ ← $\cos x = 3$

$$\cos x = \frac{1}{2}$$

$\cos x = \frac{1}{2}$ at $\frac{\pi}{3}$ and $\frac{5\pi}{3}$

←cos x can never equal 3, (look at the graph of y = cos x to see that it never goes above 1), so there are two answers to this problem.

This equation does not produce a response; it is undefined. So the two answers to the problem are $x = \frac{\pi}{3}$ and $\frac{5\pi}{3}$.

TRY IT!

1. If $\cos x = -\frac{2}{3}$ and x is in quadrant II, find the cos 2x and sin 2x.

2. Solve the equation $2\cos^2 x - 7\cos x + 3 = 0$.

Answers

1. We use the double-angle formulas for lowering powers repeatedly. Starting with the one for cosine:

$$\cos 2x = 2\cos^2 x - 1$$

$$= 2\left(-\frac{2}{3}\right)^2 - 1 = \frac{8}{9} - 1 = -\frac{1}{9}$$

To use the formula sin 2x=2sinxcosx, we need to find sin x first. We have

$$\sin x = \sqrt{1-\cos^2 x} = \sqrt{1-\left(-\frac{2}{3}\right)^2} = \frac{\sqrt{5}}{3}$$

where we have used the positive square root because sin x is positive in quadrant II. Thus,

$$\sin 2x = 2\sin x \cos x$$

$$= 2\left(\frac{\sqrt{5}}{3}\right)\left(-\frac{2}{3}\right) = -\frac{4\sqrt{5}}{9}$$

Terms You Should Know

Amplitude - The distance from the x-axis to the curve in a trigonometric graph.

Conjugate - Two binomials whose only difference is the sign of one term.

Degree - A degree is equal to $\frac{1}{360}$ of a circle.

Radian - A radian is the length of an arc of the circle, divided by the radius of the circle.

Trigonometric Identity - A trigonometric identity is a form of proof in which you use known properties of the trig functions as well as known identities of the trig functions to show that other trig identities are true.

Unit Circle - A circle of radius one with center at the origin.

Pre-Calculus

A pre-calculus course is designed to prepare a student for the study of calculus. Pre-calculus courses are often only a semester in length and are taught with trigonometry in the next semester. Pre-calculus courses may cover a variety of topics, and teachers may focus on many skills depending on whether the course is a preparation for a general calculus course or for an advanced-placement calculus course.

Pre-calculus courses usually include a review and extension of topics covered in Algebra II, as students need to have a thorough and complete understanding of all concepts from Algebra II before entering calculus. These topics include functions and operations like addition, subtraction, multiplication, division, and **composition of functions**, **rational expressions** and equations, graphing rational equations, and general work with solving equations. Also, students will spend a significant amount of time in a pre-calculus course working with logarithmic functions and the inverse of logarithmic functions.

Pre-calculus courses also typically include a discussion of **polar coordinates** and graphing polar functions. In addition, students typically encounter vectors in pre-calculus, although the work done with vectors in most pre-calculus courses is introductory.

Finally, many pre-calculus courses will include an introduction to the concept of a limit, which is of critical importance in the study of calculus.

It's important to note that pre-calculus courses, like all mathematics courses, have changed tremendously with the availability of graphing calculators and graphing software packages. Not even a decade ago, many weeks of a pre-calculus course had to be spent on exercise after exercise demonstrating how to graph complex functions by hand. Now these graphs can be rendered in seconds by entering them into a calculator. While most courses still include a unit or section on graphing functions it is no longer the focus of the course.

Example 1

Solve $x^3 - 4x^2 + x + 6 = 0$.

Recall from our study of Algebra II that we learned about something called the Fundamental Theorem of Algebra. This theorem says an equation will have exactly as many roots as its highest degree. So, in this case, we will have exactly three roots of this equation that will yield solutions. Remember that we found a list of possible rational roots (or solutions) using the Rational Root Theorem.

The **Rational Root Theorem** says that for any rational equation, all of the possible rational roots can be found by dividing each of the factors of the constant term by each of the factors of the leading coefficient. In most cases, you will see this written in a way where the factors of the constant term are labeled as *p* and the factors of the leading coefficient are labeled *q,* and you must list all of the possible combinations of $\frac{p}{q}$.

Our leading coefficient is 1 and our constant term is 6, so let's begin by making a list of all the factors of 1 and 6:

6: $\pm$ 1, $\pm$ 2, $\pm$ 3, $\pm$ 6 ←These are the values of *p.*

1: $\pm$ 1 ←These are our values for *q.*

So, if we list all the possible combinations of $\frac{p}{q}$, we get $\pm$ 1, $\pm$ 2, $\pm$ 3, $\pm$ 6. This means we could have to do eight different division problems to find the roots! (Luckily, once we find the first root, we'll often be able to solve for the second and third roots by factoring, so we'll most likely only have to divide once or twice.)

Let's start by dividing by the factor that would give us +1 as a root; this would be *x* – 1. When x-1=0, so x=1. So, we divide:

$$\begin{array}{r}
x^2-3x-2 \\
x-1\overline{)x^3-4x^2+x+6} \\
-\left(x^3-\ x^2\right) \\ \hline
-3x^2+x+6 \\
-\left(-3x^2+3x\right) \\ \hline
-2x+6 \\
-\ (-2x+2) \\ \hline
4
\end{array}$$

Since we got a remainder, $x - 1$ is not a root of the equation. Now, we need to try a root that would give us -1 as a solution, which is $x + 1$. Again, we divide; if we get no remainder, then we have a solution.

$$\begin{array}{r}
x^2-5x+6 \\
x+1\overline{)x^3-4x^2+x+6} \\
-\left(x^3+x^2\right) \\ \hline
-5x^2+x+6 \\
-\left(-5x^2-5x\right) \\ \hline
6x+6 \\
-\ (6x+6) \\ \hline
0
\end{array}$$

Fantastic! Since we got a remainder of 0, we know that $x + 1$ is a root of the equation. We also know that the quotient $\left(x^2-5x+6\right)$ is a factor of the equation. We can factor this using our Algebra I factoring skills.

$\left(x^2-5x+6\right)$

$=(x-3)(x-2)$ ←Factor (remember you can check using FOIL.)

So, we know that $x^3-4x^2+x+6=0$ can be rewritten as $(x+1)(x-3)(x-2)=0$, which we can solve easily by setting each factor equal to zero:

$x+1=0$	$x-3=0$	$x-2=0$
$x=-1$	$x=3$	$x=2$

The three roots of the equation are -1, 2, and 3.

Exercise 1

Solve $x^4+4x^3+3x^2-4x-4=0$

Recall the Fundamental Theorem of Algebra, which tells us there are at most four real solutions to this problem, because the highest degree is 4 (counting repeated zeros and complex zeros, there are exactly 4 zeros). Then we use the Rational Root Theorem to list all the possible roots of the equation. List the values of *p* and *q* so that you can list the possible roots of this equation. Values are:

p: $\pm$ 1, $\pm$2, $\pm$4

q: $\pm$ 1

So, the values of $\frac{p}{q}$ are $\pm$ 1, $\pm$2, $\pm$4.

Now, we use trial and error to find which of these is a root of the equation. In Example 1 we used long division to test the roots. Here we'll demonstrate another method of division called **synthetic division**. In synthetic division you write the coefficients of the polynomial across the top, next to the root you are dividing into them. Make sure the coefficients are listed from highest degree to lowest, and then you use 0 as a place-holder if one or more of the degrees is missing. Then, you will bring down the first value, multiply it by the root, and write it under the second number. Add the new number to the second number and then repeat the process of writing under the next number and adding until you reach the end of the row.

Watch the process here:

$$\begin{array}{r|rrrrr} & 1 & 4 & 3 & -4 & -4 \\ 1 & & 1 & 5 & 8 & 4 \\ \hline & 1 & 5 & 8 & 4 & 0 \end{array}$$

Note that the 1 in the box to the left is the root we are dividing, so since this gave us a remainder of 0 (the last number on the right), then 1 is a root of the equation so we know that (x – 1) is a factor of the equation **Think:** $x = 1$ if $x - 1 = 0$. We also know that $x^3 + 5x^2 + 8x + 4$ is a factor. When you get to this point, you can choose to rework the Rational Root Theorem with this new factor or, since we have such a short list of factors, you can simply keep trying factors in your original equation. Let's do that by trying -1.

$$\begin{array}{r|rrrrr} & 1 & 4 & 3 & -4 & -4 \\ -1 & & -1 & -3 & 0 & 4 \\ \hline & 1 & 3 & 0 & -4 & 0 \end{array}$$

We now know two of the factors of our equation, (x – 1), from the first division problem and now (x + 1).

Let's continue using synthetic division to find the remaining roots.

$$\begin{array}{r|rrrrr} & 1 & 4 & 3 & -4 & -4 \\ 2 & & 2 & 12 & 30 & 52 \\ \hline & 1 & 6 & 15 & 26 & 48 \end{array}$$

← 2 is not a root since we got a remainder of 48.

$$\begin{array}{r|rrrrr} & 1 & 4 & 3 & -4 & -4 \\ -2 & & -2 & -4 & 2 & 4 \\ \hline & 1 & 2 & -1 & -2 & 0 \end{array}$$

← -2 is a root because the remainder is 0.

$$\begin{array}{r|rrrrr} & 1 & 4 & 3 & -4 & -4 \\ 4 & & 4 & 32 & 140 & 544 \\ \hline & 1 & 8 & 35 & 136 & 540 \end{array}$$

← 4 is not a root since we got a remainder of 540.

$$\begin{array}{r|rrrrr} & 1 & 4 & 3 & -4 & -4 \\ -4 & & -4 & 0 & -12 & 64 \\ \hline & 1 & 0 & 3 & -16 & 60 \end{array}$$

← -4 also is not a root since we got a remainder of 60.

So, we have factors of (x + 1)(x – 1) and (x + 2), but the Fundamental

Theorem of Algebra told us there would be four roots. Well, there are four roots – one of these is repeated. Technically, we don't need to know which one is repeated – we simply need to list the roots as solutions. These are x = and -2, -1, and 1.

Exercise 2

Solve $\dfrac{x^5 - x^4 - 15x^3 + 25x^2 14x - 24}{(x+1)(x-1)} = 0$.

Before we begin this problem, it's important to note that if we get values of x = -1 or x = 1 as roots of the numerator, we will need to exclude them from the solution because they would make the denominator undefined (they would also cancel with the denominator terms and therefore would be excluded anyway). So, let's begin by listing the values of p and q.

p: ±1, ±2, ±3, ±4, ±6, ±8, ±12, ±24

q: ±1

So, $\frac{p}{q} =$ ±1, ±2, ±3, ±4, ±6, ±8, ±12, ±24. Also, remember that we are looking for five roots,but if 1 or -1 are roots of the numerator they will be excluded from the solution of the problem. Let's use synthetic division again.

$$\begin{array}{r|rrrrrr} & 1 & -1 & -15 & 25 & 14 & -24 \\ 1 & & 1 & 0 & -15 & 10 & 24 \\ \hline & 1 & 0 & -15 & 10 & 24 & 0 \end{array}$$

← 1 is a root, but we already know it will be excluded.

$$\begin{array}{r|rrrrrr} & 1 & -1 & -15 & 25 & 14 & -24 \\ -1 & & -1 & 2 & 13 & -38 & 24 \\ \hline & 1 & -2 & -13 & 38 & -24 & 0 \end{array}$$

← -1 is a root, but we know it must be excluded.

$$\begin{array}{r|rrrrrr} & 1 & -1 & -15 & 25 & 14 & -24 \\ 2 & & 2 & 2 & -26 & -2 & 24 \\ \hline & 1 & 1 & -13 & -1 & 12 & 0 \end{array}$$

← 2 is a root and it is also a solution.

$$\begin{array}{r|rrrrrr} & 1 & -1 & -15 & 25 & 14 & -24 \\ -2 & & -2 & 6 & 18 & -86 & 144 \\ \hline & 1 & -3 & -9 & 43 & -72 & 120 \end{array}$$

← -2 is not a root.

	1	−1	−15	25	14	−24
3		3	6	−27	−6	24
	1	2	−9	−2	8	0

← 3 is a root and it is also a solution.

Recall that we've found four of our five possible roots (assuming one of those is not a repeated root as in the example above). We're looking for one more root.

	1	−1	−15	25	14	−24
−3		−3	12	9	−102	264
	1	−4	−3	34	−88	240

←-3 is not a root.

	1	−1	−15	25	14	−24
4		4	12	−12	52	264
	1	3	−3	13	66	240

←4 is not a root.

	1	−1	−15	25	14	−24
−4		−4	20	−20	−20	24
	1	−5	5	5	−6	0

←So our final root is -4.

So, the solutions to this problem are x = -4, 2, and 3.

Exercise 3

Solve $x^3 - 19x + 30 = 0$

We've used the Fundamental Theorem of Algebra and the Rational Root Theorem to solve two equations, but there is another way to solve equations like this, and that is by graphing. While graphing is not always the most practical method, with the use of graphing software and graphing calculators, graphing as a solution method has become more prevalent. In fact, most graphing calculators and graphing software packages will include a method for finding roots (or zeroes as they are sometimes called in equations like this). A root is the place where the graph crosses the x axis. Again, the Fundamental Theorem of Algebra tells us that for a third-degree equation, this should have three roots. Let's take a look at the graph.

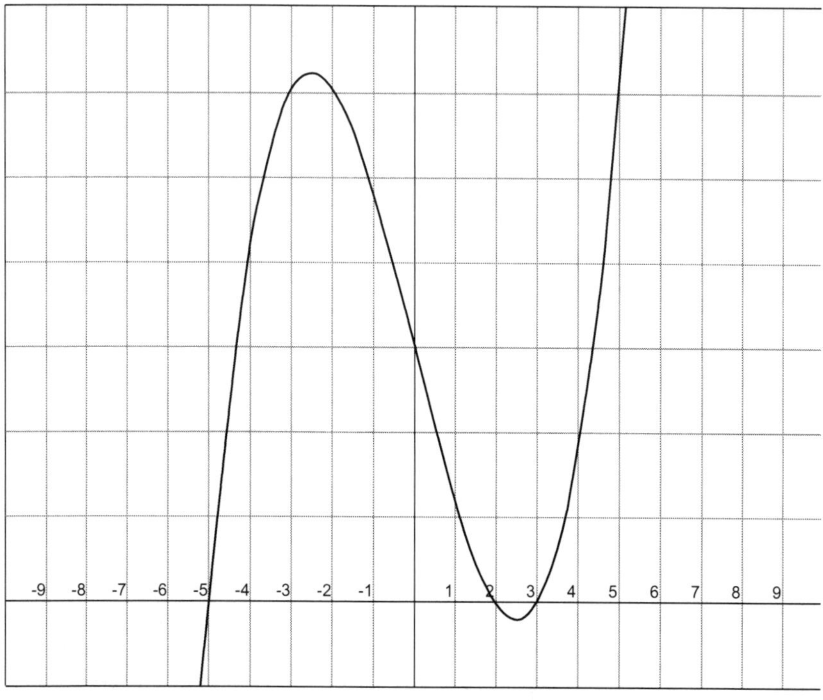

It's pretty easy to tell from the graph that the roots are -5, 2, and 3. You should check those in the equation to make sure they give you a correct solution to be sure there are no extraneous solutions. Let's do that quickly:

$x^3 - 19x + 30 = 0$

Checking x = -5 → $(-5)^3 - 19(-5) + 30 = -125 + 95 + 30 = 0$

Checking x = 2 → $(2)^3 - 19(2) + 30 = 8 - 38 + 30 = 0$

Checking x = 3 → $(3)^3 - 19(3) + 30 = 27 - 57 + 30 = 0$

So, x = -5, 2, 3

TRY IT!

1. Find the roots of the polynomial: $P(x) = 3x^3 - 30x^2 + 75x$.

2. Find all zeros of $P(x)=3x^4-2x^3-x^2-12x-4$.

Answers

1. Since 3x is a common factor, we have

$$P(x)=3x\left(x^2-10x+25\right)$$

$$P(x)=3x\left(x-5\right)^2$$

So, the roots or zeros of the function *P(x)* are x = 0 and x = 5 (with a multiplicity of 2, which means it is a repeated root).

2. Recall that roots of a polynomial are also referred to as the zeros of the polynomial. Using the Rational Zeros Theorem, we obtain the following list of possible rational zeros: ±1, ±2, ±4, $\pm\frac{1}{3}$, $\pm\frac{2}{3}$, $\pm\frac{4}{3}$. We can start by testing any of these possible rational zeros, so we'll begin with the integers.

	3	−2	−1	−12	−4
1		3	1	0	−12
	3	1	0	−12	−16

← 1 is not a root.

	3	−2	−1	−12	−4
−1		−3	5	−4	16
	3	−5	4	−16	12

← -1 is not a root.

	3	−2	−1	−12	−4
2		6	8	14	4
	3	4	7	2	0

← 2 is a root.

At this point, we know that $(x-2)$ is a factor, and $P(x)=(x-2)(3x^3+4x^2+7x+2)$. We can now shift our search for rational zeros to the polynomial $3x^3+4x^2+7x+2$, which has possible rational roots of $\pm1,\pm2,\pm\frac{1}{3},\pm\frac{2}{3}$. We've already tested ±1 (without success), and we notice that substituting a positive number in for *x*, can't result in zero, since all of the terms in the polynomial are being added. Thus, we only have three potential rational zeros to check: $-\frac{1}{3}, -\frac{2}{3}$, and -2. Trial and error leads us to the following:

$$\begin{array}{r|rrrr} -\frac{1}{3} & 3 & 4 & 7 & 2 \\ & & -1 & -1 & -2 \\ \hline & 3 & 3 & 6 & 0 \end{array} \quad \leftarrow -\frac{1}{3} \text{ is a root}$$

So, we have found that $x=2$ and $x=-\frac{1}{3}$ are zeros of the polynomial, and $P(x)$ can be factored as:

$$P(x)=(x-2)\left(x+\frac{1}{3}\right)\left(3x^2+3x+6\right)$$

$$P(x)=3(x-2)\left(x+\frac{1}{3}\right)\left(x^2+x+2\right)$$

Using the quadratic formula, we see

$x=\frac{-1\pm\sqrt{1-8}}{2}=-\frac{1}{2}\pm\frac{\sqrt{-7}}{2}$. Rewriting $\sqrt{-7}$ using the imaginary unit, we find that $x=-\frac{1}{2}\pm\frac{i\sqrt{7}}{2}$.

So, there are two real and two complex zeros of *P(x)*:

$$x=2,\ -\frac{1}{3},\ -\frac{1}{2}+\frac{i\sqrt{7}}{2},\ -\frac{1}{2}-\frac{i\sqrt{7}}{2}$$

If we are asked to find all zeros over the set of complex numbers, we would report all four zeros. If we are asked to find all real solutions or zeros, we would only report $x=2$ and $x=-\frac{1}{3}$.

Example 2

You have $1,800 to invest. The bank is offering 5% interest compounded continuously. How long will you need to leave your money invested in order to have $2,400?

Recall the formula for **exponential growth** (which is the same as the formula for compound interest when interest is compounded continuously) is $A(t) = A_0e^{kt}$ where *A*(*t*) is the amount you will have after *t* years; A_0 is the original amount you invested; and *k* is the interest rate. It's important to remember that *e* is not a variable. The value *e* is a constant, much like π. You can normally find *e* on your calculator, but if you are unsure of where to find this on your calculator you can use the fact that $e \approx 2.718$ to solve the problem.

Now that we know the values of all of our variables, let's substitute everything into the formula. Remember that we are solving for *t*, the amount of time it will take to invest the money. So we are going to substitute in \$2,400 for *A*(*t*) and \$1,800 for A_0 and 0.05 for *k*. (**Think:** 5% converted into a decimal.)

$$A(t) = A_0e^{kt}$$

$2400 = 1800e^{0.05t}$	←Substitute values.
$\frac{2400}{1800} = e^{0.05t}$	←You want to get e on one side so divide by 1800.
$\frac{4}{3} = e^{0.05t}$	←Divide and simplify.
$\ln\frac{4}{3} = \ln\ e^{0.05t}$	←To cancel the *e*, you will take the natural logarithm of both sides.
$\ln\frac{4}{3} = .05t$	←The ln and *e* cancel each other out, since they are inverse functions.
$\frac{\ln\frac{4}{3}}{0.05} = t$	←Divide both sides by 0.05.
$t \approx 5.75$ years	←Use the calculator to find ln 4/3 and to simplify.

So, this means it will take approximately 5.75 years for \$1,800 to grow to \$2,400.

Example 3

Sketch the graph of $y = \frac{2}{x-3}$. Identify any horizontal and vertical **asymptotes**.

We can always make a table of values and use that to sketch a graph. Let's pick a few values for *x* just to see if we can get an idea of what is happening with this equation.

X	$y = \frac{2}{x-3}$
-3	$-\frac{1}{3}$
-2	$-\frac{2}{5}$
-1	$-\frac{2}{5}$
0	$-\frac{2}{3}$
1	-1
2	-2
3	Undefined
4	2
5	1

If we sketch these points we appear to be getting something very strange. The two points (4, 2) and (5, 1) do not appear to be associated with the other points. In fact, we're getting two different curves, which you'll see as soon as we sketch the graph.

The other important concept that helps you see the behavior of the graph is to look at asymptotes. A graph approaches an asymptote and may even cross it once or twice (as long as it isn't a vertical asymptote), but the "ends" of the graph will never touch the asymptote. This is called the "end behavior" of the graph. Vertical asymptotes are easy to find—they occur anywhere the function is undefined. So, in this case, that can only happen when the denominator is equal to 0, so at *x* = 3. (💡 **Think:** x - 3, the denominator, ≠ 0.) So x ≠ 3.

To find horizontal asymptotes, there is a general set of rules to follow. If the numerator and the denominator have the same degree, then the asymptote is located at the ratio of the leading coefficients of the numerator and denominator. Let's look at a couple of examples:

In $y=\frac{5x^2+2}{2x^2-3}$ the horizontal asymptote would be at $y=\frac{5}{2}$; the ratio of the x^2 terms. In $y=\frac{3x^3-4}{6x-x^3}$ the horizontal asymptote would be at $y=\frac{3}{-1}=-3$.

If the numerator and denominator do not have the same degree, you must determine if the function is bottom-heavy or top-heavy. Look at which one has the higher degree to determine top- or bottom-heaviness.

If the numerator has higher degree, then the function is top-heavy and has no horizontal asymptote. If the denominator has higher degree (as in our example) then it is bottom-heavy and has an asymptote of *y* = 0.

With all this information collected, we can sketch the graph. The vertical asymptote is shown as a blue dashed line, and the horizontal asymptote (which is the *x*-axis) is shown as a yellow dashed line. Both are labeled as well.

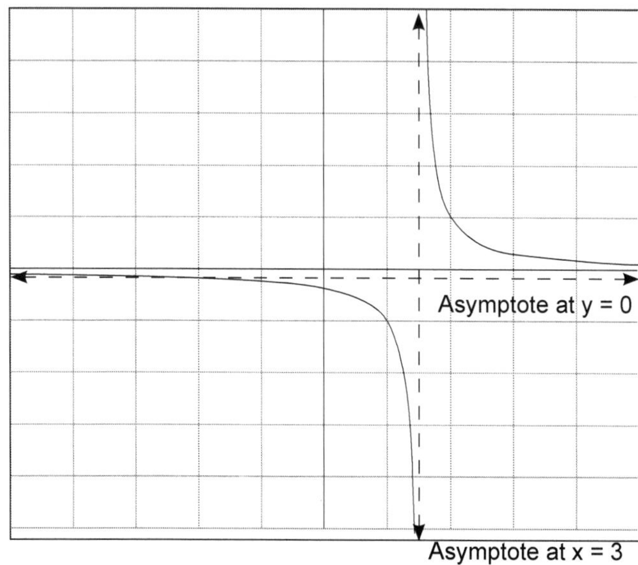

TRY IT!

In questions 1 and 2, sketch the graph of the function and identify any horizontal or vertical asymptotes that exist.

1. $f(x) = \dfrac{3x+5}{x+2}$

2. $r(x) = \dfrac{3x^2-2x-1}{2x^2+3x-2}$

Answers

1. To find the vertical asymptote, we set the denominator equal to 0 and solve:

$$f(x) = \frac{3x+5}{x+2}$$

$$x+2=0$$
$$x=-2$$

So, we have a vertical asymptote at $x = -2$. To find the horizontal asymptote, we need to compare the degree of the numerator with the degree of the denominator. Since the degrees are the same (they are both 1), a horizontal asymptote exists, and we find it by determining the ratio of the leading coefficients:

$$\text{Ratio of leading coefficients} = \frac{3}{1} = 3$$

Therefore, the equation of the horizontal asymptote is $y = 3$.

These two pieces of information will help us sketch the graph, but they are not enough on their own. We can get a better sense of the shape of the graph by finding key points on the graph such as the x- and y-intercepts. To find the x-intercepts, we set the numerator equal to zero and solve:

$$3x + 5 = 0$$
$$3x = -5$$
$$x = \frac{-5}{3}$$

So, there is a single x-intercept at $\left(\frac{-5}{3}, 0\right)$.

To find the y-intercept, substitute zero for x, and solve:

$$y = \frac{3(0) + 5}{0 + 2}$$
$$y = \frac{5}{2}$$

This means we have a y-intercept at $\left(0, \frac{5}{2}\right)$

We now have enough information to sketch the graph. First, draw the horizontal and vertical asymptotes with a dashed line. Next, plot the x- and y-intercepts.

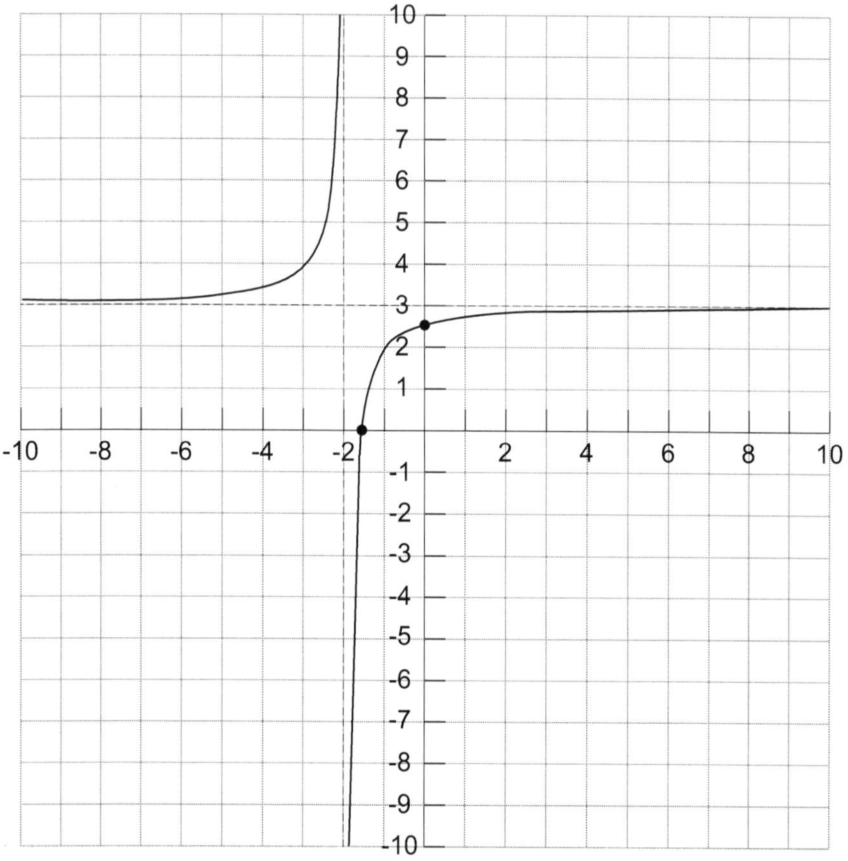

Because of the nature of asymptotes, we know that the graph will head toward positive or negative infinity as we approach x = -2 from the left. Similarly, the graph will approach y = 3 as we go further to the left. We also note that the graph can't cross the x-axis to the left of x = -2, since there are no other x-intercepts. Putting all of this information together, we sketch the graph above the horizontal asymptote and to the left of the vertical asymptote, as shown.

By a similar analysis, we recognize that the graph needs to pass through $\left(0,\frac{5}{2}\right)$ and $\left(\frac{-5}{3},0\right)$, head toward positive or negative infinity as we approach x = -2 from the right, and approach y = 3 as we go further to the right. We also note that the graph won't cross the x-axis to the right of x = -2, since there are no additional x-intercepts. This information leads us to sketch the graph below the horizontal asymptote and to the right of the vertical asymptote as shown.

2. Once again, we find the vertical asymptote by setting the denominator equal to zero and solving:

$$r(x) = \frac{3x^2 - 2x - 1}{2x^2 + 3x - 2}$$

$$2x^2 + 3x - 2 = 0$$
$$(2x - 1)(x + 2) = 0$$
$$x = \frac{1}{2},\ x = -2$$

Thus, we have two vertical asymptotes at $x = -2$ and $x = ½$. To find the horizontal asymptote, we note that both the numerator and denominator have the same degree, 2, and we find the horizontal asymptote by determining the ratio of the leading coefficients. The horizontal asymptote occurs at $y = \frac{3}{2}$. Next, we find the x-intercepts by setting the numerator equal to zero and solving:

$$3x^2 - 2x - 1 = 0$$
$$(3x + 1)(x - 1) = 0$$
$$x = \frac{-1}{3},\ x = 1$$

This means we have two x-intercepts at $\left(-\frac{1}{3}, 0\right)$ and $(1, 0)$.

To find the y-intercept, substitute zero for x and solve:

$$y = \frac{3(0) - 2(0) - 1}{2(0) + 3(0) - 2}$$
$$y = \frac{-1}{-2}$$
$$y = \frac{1}{2}$$

The y-intercept is therefore $\left(0, \frac{1}{2}\right)$.

To sketch the graph, we start by drawing the vertical and horizontal asymptotes with dashed lines, as shown. Next, plot the two x-intercepts and the y-intercept. Using a similar analysis as was done in Question 1, we can then sketch the graph as shown below.

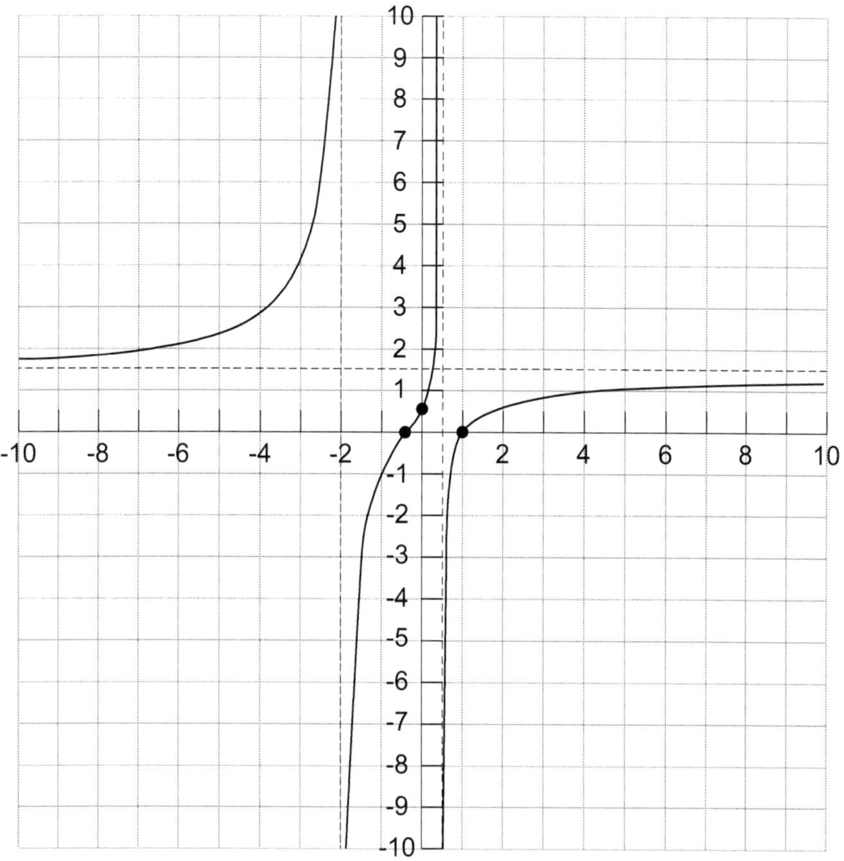

As an alternate approach, we also can choose to create a table using x values close to the vertical asymptotes to see which direction the graph is heading. Using the table, the intercepts, and our knowledge of how the graph behaves around asymptotes, we can get a very good sense of the shape of the graph.

Example 4

Decompose the fraction $\frac{1-x}{x(x+1)}$ into partial fractions.

To **decompose a fraction** means to find the two (or more) fractions that, when added together, give you the fraction with which you started. There is a simple process for doing this. First you need to factor the denominator if it is not already factored. Since our denominator is already factored, we can move on to the next step, which is to write two new fractions that use the factors of the original fraction's denominators as the denominators of the new, partial fraction's denominators.

$$\frac{1-x}{x(x+1)} \Rightarrow \frac{A}{x}+\frac{B}{x+1}$$

THINK: x(x+1)=0 so x=0 and x+1=0 or x=-1.

The third step is to rewrite these two new fractions with a common denominator.

$$\frac{A}{x}+\frac{B}{x+1}=\frac{A(x+1)+B(x)}{x(x+1)}$$

Now simplify the numerator of the new fraction.

$$\frac{A(x+1)+B(x)}{x(x+1)}-\frac{Ax+A+Bx}{x(x+1)}$$

Collect like terms in the numerator.

$$\frac{Ax+A+Bx}{x(x+1)}=\frac{Ax+Bx+A}{x(x+1)}$$

Finally, factor an *x* out of the first two terms in the numerator so you have $x(A+B)+A$. Set that equal to the original numerator, which is what we were trying to split apart.

$$x(A+B)+A=1-x$$

$$x(A+B)+A=-x+1$$

←Rearrange the right side so that the *x* termcomes first like on the left side.

Now you can set each of the sets of coefficients equal to one another so:

$A + B = -1$ and $A = 1$

Since $A = 1$, you can substitute that into $A + B = -1$:

$1 + B = -1$

$B = -2$

Now, use these values for A and B in the original statement you wrote.

$$\frac{1-x}{x(x+1)} \Rightarrow \frac{A}{x} + \frac{B}{x+1} = \frac{1}{x} - \frac{2}{x+1}$$

If you're not sure it's right, you can always multiply this back out.

There is a second method for working problems like this. You still begin the problem in the same way, using A and B as numerators and the factors of the denominator as the denominators of the two new fractions (so, our first step is the same as above).

$$\frac{1-x}{x(x+1)} \Rightarrow \frac{A}{x} + \frac{B}{x+1}$$

Next you eliminate the denominators by multiplying both sides by the least common denominator, which would be $x(x + 1)$. This will leave us with $1 - x = A(x+1) + B(x)$. Now you set the original denominator equal to 0 and you solve for the denominator, which will give you either $x = 0$ or $x = -1$. You substitute these two values into this equation and this will give you the values of A and B. We'll work these two problems side by side.

Substitute $x = 0$		Substitute $x = -1$
$1 - 0 = A(0+1) + B(0)$	←Substitution→	$1 - -1 = A(-1+1) + B(-1)$
$1 = A$	←Simplification→	$2 = -1B$ so $B = -2$

Now you substitute the values of A and B back into the first line of the problem, which gives you:

$$\frac{1-x}{x(x+1)} \Rightarrow \frac{A}{x}+\frac{B}{x+1}=\frac{1}{x}-\frac{2}{x+1}.$$

Exercise 1

Decompose the fraction $\frac{x-8}{(x+1)(x-2)}$.

$\frac{A}{x+1}+\frac{B}{x-2}$ ←Rewrite the function as two fractions, each over one factor of the denominator.

$\frac{A(x-2)+B(x+1)}{(x+1)(x-2)}$ ←Use these two new fractions and get a common denominator.

$\frac{Ax-2A+Bx+B}{(x+1)(x-2)}$ ←Distribute *A* and *B* in the numerator.

$\frac{Ax+Bx-2A+B}{(x+1)(x-2)}$ ←Collect like terms.

Now, we know the following must be true:

$Ax+Bx-2A+B=x-8$ ←$x-8$ is the numerator of our original equation,

So that information tells us that the *x* terms on the two sides of the equal sign must be equal and the constant terms on each side of the equal sign must be equal.

$$Ax+Bx=x$$
$$-2A+B=-8$$

Now we'll simplify these and solve for *A* and *B* by solving the system of equations.

$$\begin{aligned} A+B&=1 \\ -2A+B&=-8 \end{aligned} \quad \Rightarrow \quad \begin{aligned} -A-B&=-1 \\ \underline{-2A+B}&\underline{=-8} \\ -3A\quad&=-9 \\ A\quad&=\ 3 \end{aligned}$$

Use the system of equations, if A = 3, then B = -2. Now, we just substitute that back into the first fraction in our decomposition.

$$\frac{A}{x+1}+\frac{B}{x-2}=\frac{3}{x+1}-\frac{2}{x-2}$$

TRY IT!

1. Find the partial fraction decomposition of $\frac{5x+7}{x^3+2x^2-x-2}$.

2. Find the partial fraction decomposition of $\frac{2x^2-x+4}{x^3+4x}$.

Answers

1. The denominator factors as follows:

$$x^3+2x^2-x-2=x^2(x+2)-(x+2)=(x^2-1)(x+2)=(x-1)(x+1)(x+2)$$

This gives us the partial fraction decomposition:

$$\frac{5x+7}{x^3+2x^2-x-2}=\frac{A}{(x-1)}+\frac{B}{x+1}+\frac{C}{x+2}$$

Multiply each side by the common denominator, (x-1)(x+1)(x+2), we get:

$$5x+7=A(x+1)(x+2)+B(x-1)(x+2)+C(x-1)(x+1)$$
$$=A(x^2+3x+2)+B(x^2+x-2)+C(x-1)(x+1)$$
$$=(A+B+C)x^2+(3A+B)x+(2A-2B-C)$$

If the two polynomials are equal, then their coefficients are equal. Since 5x+7 has no x^2 terms, we have A+B+C=0. Similarly, by comparing the coefficients of x, we see that 3A+B=5, and by comparing constant terms, we get 2A-2B-C=7. This leads us to the following system:

$$\begin{cases} A+B+C=0 \\ 3A+B=5 \\ 2A-2B-C=7 \end{cases}$$

THINK: Equation 1: coefficients of x^2; Equation 2: coefficient of x; Equation 3: constant coefficients.

We can solve this system manually or by using a graphing calculator. In doing so, we see that A=2, B=-1 and C=-1.

So the partial decomposition is:

$$\frac{5x+7}{x^3+2x^2-x-2}=\frac{2}{x-1}+\frac{-1}{x+1}+\frac{-1}{x+2}$$

2. Since $x^3+4x=x(x^2+4)$ can't be factored further, we write:

$$\frac{2x^2-x+4}{x^3+4x}=\frac{A}{x}+\frac{Bx+C}{x^2+4}$$

Multiplying by $x(x^2+4)$, we get:

$$2x^2-x+4=A(x^2+4)+(Bx+C)x$$
$$=(A+B)x^2+Cx+4A$$

Equating coefficients gives us the equations:

$$\begin{cases} A+B=2 \\ C=-1 \\ 4A=4 \end{cases}$$

THINK: Equation 1: coefficients of x^2; Equation 2: coefficient of x; Equation 3: constant coefficients.

So A=1, B=1, and C=-1. The required partial fraction decomposition is

$$\frac{2x^2-x+4}{x^3+4x}=\frac{1}{x}+\frac{x-1}{x^2+4}$$

Example 5

Find the position vector associated with $\vec{v} = \langle (1,2),(3,-5) \rangle$.

Before we can do this problem, we need to know what a vector and position vector are. A **vector** is a line segment that has both an angle and direction. This means a vector, when drawn on a graph, will have its end-point at the first point, (1, 2), and will be pointed in the direction of the second point, (3, -5), in this case.

A **position vector** is a vector that is the same length as a given vector but has its initial point at the origin. To find a position vector, let the first point of the new vector be (0, 0), which is the origin. Then subtract the values in the first ordered pair from the values in the second ordered pair, and this gives the new second coordinate. So, this would give us the vector

$\langle (0,0),(3-1,-5-2) \rangle$

If we simplify, this gives us the vector $\vec{v}_p = \langle (0,0),(2,-7) \rangle$.

Exercise 1

Given $\vec{u} = \langle 2,-1 \rangle$ and $\vec{v} = \langle 3,5 \rangle$. Find $2\vec{u} - \vec{v}$

To find $2\vec{u} - \vec{v}$ you will need to multiply each value in vector $\vec{u} = \langle 2,-1 \rangle$ by 2 and then add this to vector $\vec{v} = \langle 3,5 \rangle$. This gives us 2(2) – 3 for the first value, which is 1, and

2(-1) – 5 for the second value, which is -7. We write this as a vector $2\vec{u} - \vec{v} = \langle 1,-7 \rangle$.

Exercise 2

Find the length of vector $\vec{x} = \langle (3,-1),(5,8) \rangle$.

The length of a vector is represented by putting the variable in symbols that resemble an absolute value symbol. To find the length of the vector you use the distance formula from geometry on the two points that form the vector. This is:

$$|\vec{x}| = \sqrt{(x_2 - x_1)^2 + (y_2 - y_1)^2}$$

$$|\vec{x}| = \sqrt{(5-3)^2 + (8-(-1))^2} = \sqrt{2^2 + 9^2} = \sqrt{4+81} = \sqrt{85}$$

Exercise 3

Given that $\vec{u} = \langle 4, -2 \rangle$ and $\vec{v} = \langle -3, 1 \rangle$, find $\vec{u} + 3\vec{v}$.

To find $\vec{u} + 3\vec{v}$, find each of the values:

4 + 3(-3) = 4 + -9 = -5

and

-2 + 3(1) = -2 + 3 = 1

So $\vec{u} + 3\vec{v} = \langle -5, 1 \rangle$.

TRY IT!

1. Find the position vector associated with $\vec{v} = \langle (.5, -3), (13, -2) \rangle$

2. Find the position vector associated with $\vec{v} = \langle (.5, -3), (13, -2) \rangle$

Answers

1. $\vec{v}_p = \langle (-2 - -3), (13 - .5) \rangle = \langle (0,0), (1, 12.5) \rangle$

2. $\vec{v}_p = \langle (-1 - -5), (10 - 4) \rangle = \langle (0,0), (4, 6) \rangle$

Exercise 1

Find the half-life of a radioactive element if it takes 2 years for a 20g sample to decay to 14g.

Again, we will use the exponential growth formula, but this time we are trying to find a value for *k*, the rate of decay of the element. We'll then use *k* to find the half-life of the element—the time it takes to decay to half the original amount.

So, we start with the formula $A(t) = A_0 e^{kt}$ and substitute 14 for *A*(*t*), 20 for A_0 and 2 for *t*.

$14 = 20e^{2k}$

$\frac{14}{20} = e^{2k}$ ←Divide both sides by 20 to get *e* by itself.

$0.7 = e^{2k}$ ←Simplify.

$\ln 0.7 = \ln e^{2k}$ ←Take the natural log of both sides to cancel *e*.

$\ln 0.7 = 2k$ ←Simplify.

$\frac{\ln 0.7}{2} = k$ ←Divide by 2 to solve for *k*.

$-0.178 \approx k$ ←Simplify using a calculator.

Now, we'll use this value for *k* to find the value for *t* that corresponds to the half-life. Regardless of the amount with which we start, (whether 20 grams or 1 gram), the half-life will be the same. So, to keep the calculations simple, we'll use 2 grams as our initial amount (A_0), and find out how long it takes until that amount decays to 1 gram – half its original amount. Thus, we'll substitute $A_0 = 2$, $A(t) = 1$, and $k = -0.178$ into our original equation and solve for *t*.

$1 = 2e^{-0.178t}$ ←Substitute.

$\frac{1}{2} = e^{-0.178t}$ ←Divide by 2.

$\ln 0.5 = \ln e^{-0.178t}$ ← Take the natural log of both sides.

$\ln 0.5 = -0.178t$ ← Simplify.

$\frac{\ln 0.5}{-0.178} = t$ ← Divide.

$t \approx 3.894$ years

Exercise 2

The number of students infected with the flu at the local high school after *t* days from the beginning of a flu outbreak is given by the function:

$$I(t) = \frac{950}{1+70e^{-0.3t}}$$

The school closes if more than 225 students become infected with the flu. After how many days is this most likely to occur?

In this problem, we want to find out when the number of infected students will reach 225, so we will substitute 225 for $I(t)$.

$225 = \frac{950}{1+70e^{-0.3t}}$ ←Substitute.

$225\left(1+70e^{-0.3t}\right) = 950$ ←Cross-multiply to eliminate the fraction.

$225 + 15750e^{-0.3t} = 950$ ←Distribute the 225 through the parentheses.

$15750e^{-0.3t} = 950 - 225$ ←Subtract 225 from both sides.

$15750e^{-0.3t} = 725$ ←Simplify.

$e^{-0.3t} = 0.04603$ ←Divide both sides by 15,750.

$\ln e^{-0.3t} = \ln 0.04603$ ←Take the natural log of both sides.

$-0.3t = \ln 0.04603$ ←Natural log and *e* cancel one another out.

$t = \frac{\ln 0.04603}{-0.3} \approx 10.26 \text{ days}$ ←Simplify using a calculator.

Exercise 3

You know that you need to have \$8,000 saved to attend the local community college when you graduate from high school. By the time you reach your sophomore year of high school, you have saved \$6,000 from allowance, odd jobs, and working after school waiting tables. If you invest this money at 7.5% compounded continuously, how long will it take for you to reach your goal of \$8,000?

This problem uses our exponential growth equation again. Let's substitute in the values and solve for t.

$$A(t) = A_0 e^{kt}$$

$8000 = 6000e^{0.075t}$ ←Substitute in values for $A(t)$, A_0, and k.

$\frac{8000}{6000} = e^{0.075t}$ ←Divide both sides by 6000.

$\frac{4}{3} = e^{0.075t}$ ←Simplify.

$\ln\frac{4}{3} = \ln e^{0.075t}$ ←Take the natural log of both sides.

$\ln\frac{4}{3} = 0.075t$ ←Natural log and e cancel one another out.

$\frac{\ln\frac{4}{3}}{0.075} = t$ ←Divide both sides by 0.075.

$t \approx 3.84$ years ←Simplify.

To reach your savings goal of $8,000, you will need to leave your initial investment in the bank for approximately 3.84 years at the rate of 7.5%.

Exercise 4

Sketch the graph of $y = \frac{x^3}{x-1}$. Identify the horizontal and vertical asymptotes, if they exist.

Let's start with the asymptotes. Recall that vertical asymptotes occur anywhere the function is undefined. In this case, that would occur when the denominator equals 0. **THINK:** We quickly solve $x - 1 = 0$, for x.) This function will have a vertical asymptote at $x = 1$.

To find horizontal asymptotes, we determine if the numerator and denominator have the same degree (the largest exponent in each is the same). In this case, the numerator has a degree of 3 and the denominator has a degree of 1 so this function would be called "top-heavy." A top-heavy function has no horizontal asymptote.

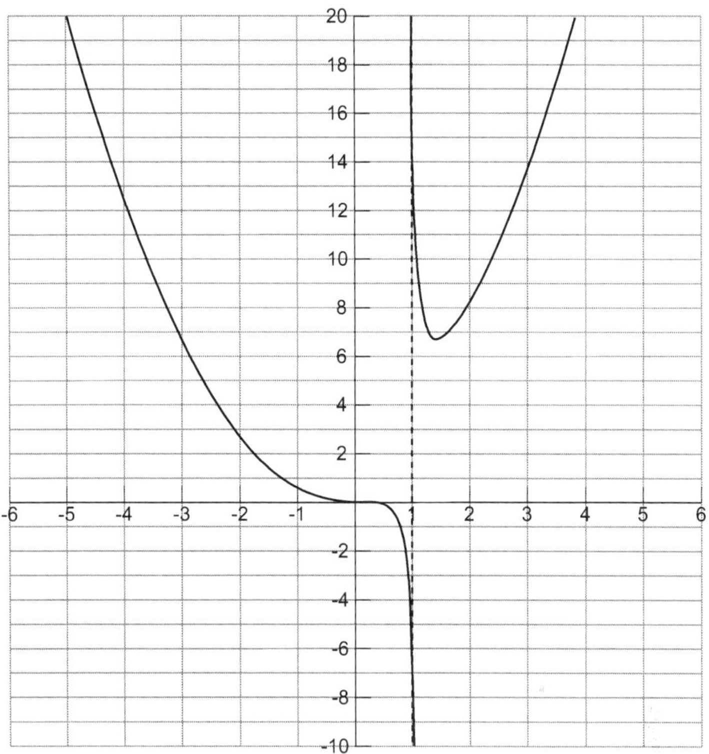

Now, to sketch the graph we can either use a graphing calculator, graphing software, or quickly find a few points by making a table of *x* values and using those to find *y*. We'll use graphing software.

Exercise 5

Use a graph to identify the solution of $3^{x-1} = 27$.

In this situation, we need to find out where the two lines $y = 3^{x-1}$ and $y = 27$ intersect. You might recall that this is the same way we found the solution to a system of equations graphically—in fact it is exactly the same process. (See page 72 for a review of how to solve a system of equations graphically.)

You can do this by making a table of values for $y = 3^{x-1}$ or by using a graphing calculator or graphing software. Again, we'll use graphing software to quickly sketch the two lines. Most graphing calculators and graphing software packages also include a calculation that finds the intersection point of two functions, so check the manual for your calculator or software package to see how to use this feature.

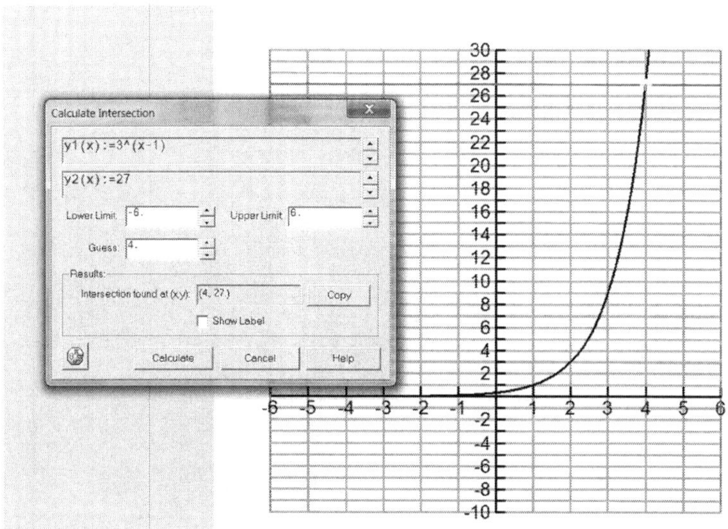

You can see from the figure that we quickly graphed the line and used the built-in intersection finder to find the solution to this equation, which is $x = 4$.

Exercise 6

Solve $2\ln\sqrt{x} = 6$ graphically.

Again, this problem would have been quite challenging before graphing calculators and graphing software became readily available, but now it is as simple as graphing $y = 2\ln\sqrt{x}$ and $y = 6$ and finding the intersection of the two graphs, either by looking at the graph and estimating or by using the calculator's or software's built-in intersection calculation method.

Using the software's intersection finder, we find out that the two functions intersect at x = 403.43 and y = 6, so the solution to the equation is x = 403.43.

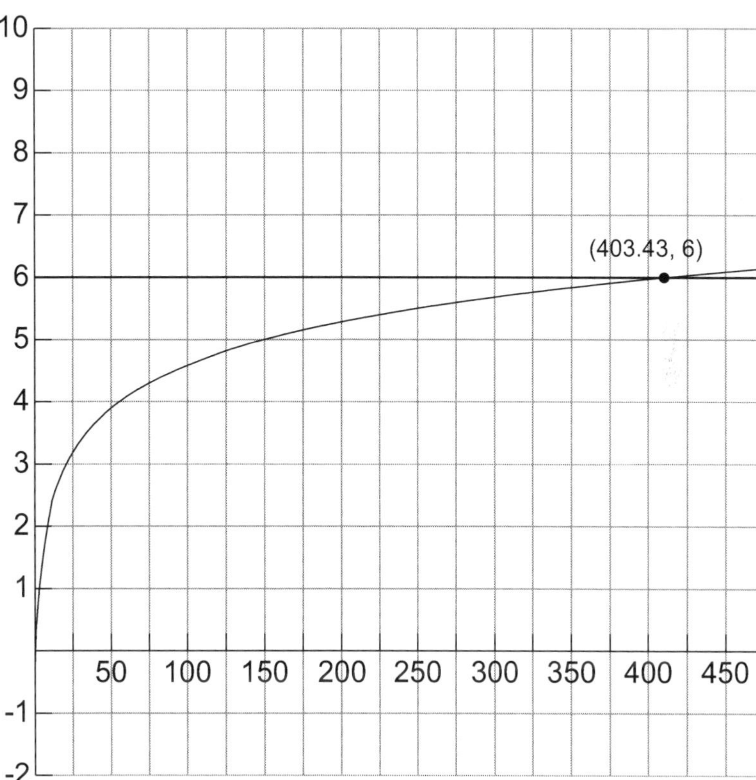

Exercise 7

Given the function $f(x) = 2x^2 - 1$, find and simplify $\frac{f(x+h) - f(x)}{h}$.

This is a particularly important example in a pre-calculus course as you prepare for the concept of a limit and using the limit to find a derivative in calculus, as the definition of a derivative involves the value of $\frac{f(x+h) - f(x)}{h}$. To complete this problem, we need to evaluate the two functions in the numerator using our given function. So, this becomes:

$\frac{\left[2(x+h)^2-1\right]-\left[2x^2-1\right]}{h}$ ←Substitute $x + h$ and x into the function.

$\frac{\left[2(x+h)(x+h)-1\right]-\left[2x^2-1\right]}{h}$ ←Write out $(x+h)^2$.

$\frac{\left[2\left(x^2+2hx+h^2\right)-1\right]-\left[2x^2-1\right]}{h}$ ←Multiply using FOIL.

$\frac{\left[2x^2+4hx+2h^2-1\right]-2x^2+1}{h}$ ←Distribute 2 through the parentheses and -1 through the bracket.

$\frac{2h^2+4hx}{h}$ ←Simplify by combining like terms.

$\frac{2h(h+2x)}{h}$ ←Factor the GCF in the numerator.

$2(h+2x)$ ←Cancel common factors.

Exercise 8

Given $f(x)=x^3+2$ and $g(x)=\frac{2}{x-1}$, find $(g\circ f)(x)$.

Recall that the symbol $\circ$ means to compute the composition of functions. In this case we are going to input the value of $f(x)$ into $g(x)$ and then we will simplify.

$(g\circ f)(x)=\frac{2}{x^3+2-1}$. Now we simplify and get $(g\circ f)(x)=\frac{2}{x^3+1}$.

TRY IT!

1. A fossil was found to contain 2.4% of its original amount of Carbon-14. If the half-life of Carbon-14 is 5,700 years, approximately how old is the fossil?

Solution:

We start with the exponential decay equation: $A(t)=A_0e^{kt}$. Our first task is to find *k*, the decay constant, by using the information we have about the half-life of Carbon-14. We know that it takes 5,700 years for Carbon-14 to decay to half its original amount. In other words, the amount of Carbon-14 left after 5,700 years is $\frac{1}{2}A_0$, and we can substitute this information into our equation as follows:

$\frac{1}{2}A_0 = A_0e^{5700k}$ ← Substitute.

$\frac{1}{2} = e^{5700k}$ ← Divide both sides by A_0.

$\ln\left(\frac{1}{2}\right) = 5700k$ ← Take the natural log of both sides.

$\frac{\ln\left(\frac{1}{2}\right)}{5700} = k$ ← Divide both sides by 5700.

$k \approx -.0001216$ ← Simplify and round.

Now that we've found *k*, we can substitute it into our original equation in order to find *t*. We know that the amount of Carbon-14 present now is 2.4% of the original amount, so $A(t) = .024(A_0)$.

$.024(A_0) = A_0e^{-.0001216t}$ ← Substitute.

$.024 = e^{-.0001216t}$ ← Divide both sides by A_0.

$\ln .024 = -.0001216t$ ← Take the natural log of both sides.

$\frac{\ln .024}{-.0001216} = t$ ← Divide both sides by *k*.

$t \approx 30{,}672$ years ← Simplify and round.

If you had chosen to hold off on rounding your work until the very last calculation, you would have found the answer to be approximately 30,671 years.

Terms You Should Know

Asymptotes - Horizontal asymptotes are horizontal lines that the graph of the function approaches as x tends to $+\infty$ or $-\infty$. Vertical asymptotes are vertical lines near which the function grows without bound.

Composite Functions - The application of one function to the results of another.

Decompose a Fraction - To decompose a fraction means to find the two (or more) fractions that when added together give you the fraction with which you started.

Exponential Growth - Occurs when the growth rate of a mathematical function is proportional to the functions current value.

Polar Coordinates - The polar coordinate system is a two-dimensional coordinate system in which each point on a plane is determined by a distance from a fixed point and an angle from a fixed direction.

Position Vector - A position vector is a vector that is the same length as a given vector but has its initial point at the origin.

Rational Root Theorem - The Rational Root Theorem says that for any rational equation, all of the possible roots can be found by dividing each of the factors of the constant term by each of the factors of the leading coefficient.

Vector - A vector is a line segment that has both angle and direction.

Calculus

The study of calculus typically is reserved for the last year of a student's high school academic program. Calculus may be studied as a single course or in two or more courses. The advanced-placement calculus program includes two separate programs of study, each accepted by different colleges and universities. If you are considering taking an advanced-placement calculus course, it is important that you review the requirements of colleges you are considering to determine if they will accept credit for the course offered by your school.

Calculus involves studying the way objects behave in the natural world. There are a multitude of calculus applications in economics, finance, physics, and most other branches of science. The development of the field of calculus is generally attributed to two individuals—Sir Isaac Newton and Gottfried Leibniz. In the 18th century both Newton and Leibniz separately extended current mathematics knowledge of the time into what we now call calculus. In calculus we use something called Leibniz Notation in honor of the contributions of Gottfried Leibniz.

There are essentially two concepts covered in a calculus course: the **derivative** and the **integral**. Both of these are based on the concept of a limit, which we began to study in the pre-calculus section, when we discussed finding horizontal asymptotes to a curve. An asymptote is roughly related to a limit as we'll see a bit later.

The **derivative** is an attempt to describe the motion of a moving object. You will hear the derivative called "the tangent line to the curve." There are a variety of applications involving the derivative, but some of the most popular involve determining the behavior of objects in relation to gravity, which was directly related to Newton's work.

The second concept studied in a calculus course is the **integral**. The derivative and the integral are inverse functions of one another, so the integral is sometimes called the anti-derivative. The study of the integral involves studying methods for finding the area under a curve.

A solid understanding of calculus is critical to work in the sciences and engineering. However, calculus often is a required course for those students entering professions in medicine, business, and architecture.

Example 1

Find $\lim_{x\to\infty} f(x) = \dfrac{3x^2+2x-5}{5x^2-x+7}$.

This problem would be read as, "Find the limit of f(x) as *x* goes to infinity, where f(x) equals the quantity 3*x* squared plus 2*x* minus 5, divided by the quantity 5*x* squared minus *x* plus 7." When you find the limit as *x* approaches infinity, you are trying to determine what happens as *x* gets larger. In general, finding limits as *x* goes to infinity with the same degree follows the same rules as those for horizontal asymptotes from pre-calculus courses. Recall that these rules say if the numerator and the denominator have the same degree, then the asymptote is located at the ratio of the leading coefficients of the numerator and denominator. Let's look at a couple of examples:

In $y = \dfrac{5x^2+2}{2x^2-3}$ the horizontal asymptote would be at $y = \dfrac{5}{2}$ — the ratio of the x^2 terms. In $y = \dfrac{3x^3-4}{6x-x^3}$ the horizontal asymptote would be at $y = \dfrac{3}{-1} = -3$.

If the numerator and denominator do not have the same degree, then you need to determine if the function is "bottom-heavy" or "top-heavy." Again, you look at which has a higher degree. If the numerator has higher degree, then the function is "top heavy" and has no horizontal asymptote. If the denominator has higher degree (as in our example), then it is "bottom-heavy" and an asymptote of *y* = 0.

So, in this particular example, the limit would be $\frac{3}{5}$.

You can also find limits by looking at a graph and observing the behavior of the graph. Obviously, you can't look at a graph that extends to infinity, but you can determine that a graph develops a pattern (such as in this case approaching the horizontal asymptote $y = \frac{3}{5}$ and never crossing it). The definition says: A function has a limit as x approaches a real number c, if and only if the limit as the function approaches c from the left and the limit as the function approaches c from the right both exist and are equal. Many functions will not have a limit.

So, if a function has a change in behavior that occurs at the point at which you are trying to take the limit, the function's limit may not exist.

Exercise 1

Find $\lim\limits_{x \to \infty}\left(5 + \frac{1}{x}\right)$.

Sums can be split up as separate limits. This would give us:

$$\lim_{x \to \infty}\left(5 + \frac{1}{x}\right) = \lim_{x \to \infty} 5 + \lim_{x \to \infty}\left(\frac{1}{x}\right)$$

The limit of a constant is the value of the constant, so the first limit is equal to 5.

Recall that to determine limits of rational functions as x approaches infinity, we determine whether the function is top-heavy, in which case no limit exists, or if the function is bottom-heavy, in which case the limit would be 0. We understand this concept because as the number in the denominator gets larger and larger (and goes closer to infinity) then the value of the function will be smaller and smaller and therefore will approach 0. The function $\frac{1}{x}$ is bottom-heavy so the value of this would be 0. Therefore:

$$\lim_{x \to \infty}\left(5 + \frac{1}{x}\right) = \lim_{x \to \infty} 5 + \lim_{x \to \infty}\left(\frac{1}{x}\right) = 5 + 0 = 5$$

Exercise 2

Find $\lim_{x\to\infty}\frac{3x^3+2}{2x}$.

This function is top-heavy since the degree of the numerator is 3 and the degree of the denominator is 1. This means the numerator will get larger faster than the denominator so the value of the rational expression will get larger and larger. Therefore, no limit exists.

Exercise 3

Find $\lim_{x\to 4}\frac{x^2}{8}$.

In this case, as *x* goes to 4, the limit will approach the value of the function at *x* = 4 so we can simply substitute 4 in for *x*.

$$\lim_{x\to 4}\frac{x^2}{8}=\frac{4^2}{8}=\frac{16}{8}=2$$

TRY IT!

1. Evaluate the following: $\lim_{x\to 5}(2x^2-3x+4)$

2. Evaluate the following: $\lim_{x\to -2}\frac{x^3+2x^2-1}{5-3x}$

3. Evaluate the following: $\lim_{x\to 0}\frac{(3+h)^2-9}{x}$

Answers

1. $\lim_{x\to 5}(2x^2-3x+4)=\lim_{x\to 5}(2x^2)-\lim_{x\to 5}(3x)+\lim_{x\to 5}(4)$
   $=2\lim_{x\to 5}x^2-3\lim_{x\to 5}x+\lim_{x\to 5}4$
   $=2(5)^2-3(5)+4$
   $=39$

We can break the limit into the sum of its parts and then substitute the value 5 for *x*.

$$2.\ \lim_{x\to -2}\frac{x^3+2x^2-1}{5-3x}=\frac{\lim\limits_{x\to -2}(x^3+2x^2-1)}{\lim\limits_{x\to -2}(5-3x)}$$

$$=\frac{\lim\limits_{x\to -2}x^3+\lim\limits_{x\to -2}2x^2-\lim\limits_{x\to -2}1}{\lim\limits_{x\to -2}5-\lim\limits_{x\to -2}3x}$$

$$=\frac{\lim\limits_{x\to -2}x^3+2\lim\limits_{x\to -2}x^2-\lim\limits_{x\to -2}1}{\lim\limits_{x\to -2}5-3\lim\limits_{x\to -2}x}$$

$$=\frac{(-2)^3+2(-2)^2-1}{5-3(-2)}$$

$$=\frac{-1}{11}$$

As in number 1, we can break the limit into the sum of its parts and then substitute the value -2 for *x* to evaluate the limit of the function.

$$3.\ \lim_{x\to 0}\frac{(3+h)^2-9}{x}=\lim_{x\to 0}\frac{(9+6x+x^2)-9}{x}=\lim_{x\to 0}\frac{6x+x}{x}=\lim_{x\to 0}(6+h)=6$$

We begin this problem by expanding the $(3+h)^2$ and combining like terms to reduce the rational expression. We can not substitute the 0 in for x as a first step, as that would put a 0 in the denominator.

Example 2

Find the derivative $f'(x)$ of $f(x)=\frac{1}{x}$ using the definition of a derivative.

In pre-calculus you often work with advanced operations on functions, and most likely a pre-calculus course will include one or more problems that ask you to find something that looks very similar to $\frac{f(x+h)-f(x)}{h}$. This rather complex looking fraction is actually the key component of the definition of a derivative. The derivative is typically represented by the symbol $f'(x)$ which is read "f prime."

To find the derivative using the definition you must find $f'(x) = \lim_{h \to 0} \frac{f(x+h) - f(x)}{h}$. Let's see how this works to find the derivative of $f(x) = \frac{1}{x}$. First we'll use $f(x)$ in the formula for the limit.

$$f'(x) = \lim_{h \to 0} \frac{f(x+h) - f(x)}{h}$$

$$f'(\frac{1}{x}) = \lim_{h \to 0} \frac{\frac{1}{x+h} - \frac{1}{x}}{h}$$ ←Substitute.

$$f'(\frac{1}{x}) = \lim_{h \to 0} \frac{\frac{1x - 1(x+h)}{x(x+h)}}{h}$$ ←Find a common denominator in the numerator.

Now you want to continue simplifying the fraction.

$$f'(\frac{1}{x}) = \lim_{h \to 0} \frac{\frac{1x - x - h}{x(x+h)}}{h}$$ ←Distribute.

$$f'(\frac{1}{x}) = \lim_{h \to 0} \frac{\frac{-h}{x(x+h)}}{h}$$ ←Combine like terms.

$$f'(\frac{1}{x}) = \lim_{h \to 0} \frac{-h}{x(x+h)} \cdot \frac{1}{h}$$ ←Dividing by h is the same as multiplying by $\frac{1}{h}$.

$$f'(\frac{1}{x}) = \lim_{h \to 0} \frac{-1}{x(x+h)}$$ ←Simplify.

Now that you have simplified the expression, you can determine that as h gets closer and closer to 0 (or even becomes 0) and you substitute that in for h, this function will approach $-\frac{1}{x^2}$, which is its derivative.

If we had to work through this definition to find every derivative, it would become a very tedious process. Fortunately, most calculus textbooks include a list of commonly used derivatives and tools you can use to find derivatives quickly. Several good listings of commonly used derivatives can also be downloaded off the Internet by doing a search for "Calculus Quick Reference" or "Calculus Reference Sheet." We've also included a handy quick reference sheet in the back of this book.

Example 3

Find the derivative of $f(x)=x^5\cdot x^{11}$ using the product rule.

This problem tells us specifically to use the product rule, however, you can find the derivative much more easily by multiplying $f(x)=x^5\cdot x^{11}=x^{16}$ and then taking the derivative (if you look at the reference sheet you'll see that the derivative can be found using the formula $f'(x)=nx^{n-1}$ for any value of n). Given this formula, we can determine that the derivative would be $f'(x)=16x^{16-1}=16x^{15}$.

However, we're supposed to use the product rule for this problem, so let's see how that would work. If you look at your quick reference sheet, you'll see that the product rule says for any two functions $f(x)\text{ and }g(x)$, $(f\cdot g)'(x)=f'(x)\cdot g(x)+f(x)\cdot g'(x)$. Many calculus teachers can recite this quickly (because they've been saying it for years) as "derivative of the first times the second plus first times derivative of the second." If you say the product rule to yourself just like this every time you use it, you too can remember the product rule just like a good calculus teacher.

So let's take a look at how that product rule applies to this problem. In this case we would need to define $f(x)\text{ and }g(x)$ as $f(x)=x^5\text{ and }g(x)=x^{11}$. So, we can say if:

$f(x)=x^5\text{ and }g(x)=x^{11}$ and $(f\cdot g)'(x)=f'(x)\cdot g(x)+f(x)\cdot g'(x)$ then

$(f\cdot g)'(x)=5x^4\cdot x^{11}+x^5\cdot 11x^{10}$ ←Substitute the values of into the formula for the Product Rule.

$f(x)$, $g(x)$, $f'(x)$, and $g'(x)$

$(f \cdot g)'(x) = 5x^{15} + 11x^{15}$ ←Simplify using rules of exponents.

$(f \cdot g)'(x) = 16x^{15}$ ←Combine like terms.

Note: We got the same answer when we simply multiplied the exponents and took the derivative, so either method works fine.

Example 4

Find the derivative of $y = \frac{5x}{x+1}$ using the quotient rule.

Just as in the problem above, there are two ways to approach this problem. The first is to rewrite this problem using negative exponents and then use the product rule, however, a more effective approach is to use the quotient rule (which is covered in your quick reference sheet, which we'll demonstrate below). The quotient rule for any two functions $f(x)$ and $g(x)$ says $\left(\frac{f}{g}\right)'(x) = \frac{f'(x) \cdot g(x) - f(x) \cdot g'(x)}{(g(x))^2}$.

You'll note that the numerator of the quotient rule is actually the product rule only with a minus sign instead of a plus sign, so if you remember that, all you have to remember to keep the quotient rule straight is to put that numerator over $(g(x))^2$. So let's define $f(x)$ and $g(x)$ for this problem and use those in the quotient rule.

$f(x) = 5x$ and $g(x) = x + 1$

$\left(\frac{f}{g}\right)'(x) = \frac{f'(x) \cdot g(x) - f(x) \cdot g'(x)}{(g(x))^2}$ ←Quotient Rule.

$\left(\frac{f}{g}\right)'(x) = \frac{5(x+1) - 5x \cdot 1}{(x+1)^2}$ ←Substitute the values of into the formula for the Quotient Rule.

$f(x)$, $g(x)$, $f'(x)$, and $g'(x)$

$\left(\frac{f}{g}\right)'(x)=\frac{5x+5-5x}{(x+1)^2}$ ←Simplify.

$\left(\frac{f}{g}\right)'(x)=\frac{5}{(x+1)^2}$ ←Combine like terms.

Exercise 1

Find the derivative $f'(x)$ of $f(x)=x^2$ using the definition of a derivative.

Recall the definition of a derivative is $f'(x)=\lim_{h\to 0}\frac{f(x+h)-f(x)}{h}$. Substitute in the values in the definition.

$f'(x)=\lim_{h\to 0}\frac{(x+h)^2-(x)^2}{h}$ ←Substitute the values in the definition of $f'(x)$.

$f'(x)=\lim_{h\to 0}\frac{(x+h)(x+h)-(x)^2}{h}$ ←Write out $(x+h)^2$.

$f'(x)=\lim_{h\to 0}\frac{(x^2+2hx+h^2)-(x)^2}{h}$ ←Multiply using FOIL.

$f'(x)=\lim_{h\to 0}\frac{2hx+h^2}{h}$ ←Combine like terms.

$f'(x)=\lim_{h\to 0}\frac{h(2x+h)}{h}$ ←Factor out the GCF.

$f'(x)=\lim_{h\to 0}(2x+h)$ ←Cancel h in numerator and denominator.

$f'(x)=2x$ ←Evaluate limit at $h = 0$.

Exercise 2

Find $f'(x)$ if $f(x)=2x+1$.

If you take a look at your reference sheet, you'll see that $f'(cx)=c$. In this case c = 2. So, $f'(x)=2$ (the derivative of 1 is 0 so you do not need a derivative of the second term).

Exercise 3

Find $f'(x)$ if $f(x)=3x^4+5x^3-9x^2+6x-1$.

To find the derivative of a sum (or difference), you simply find the derivative of each of the individual terms. So, in this case, we'll use the derivative rule $f'(x^n)=nx^{n-1}$ on each term and the derivate rule $f'(cx)=c$ on the term 6*x*.

For $f(x)=3x^4+5x^3-9x^2+6x-1$

$f'(x)=3\cdot 4x^{4-1}+5\cdot 3x^{3-1}-9\cdot 2x^{2-1}+6$ ←Apply the derivate rules from above.

$f'(x)=12x^3+15x^2-18x+6$ ←Simplify.

Exercise 4

Find the derivative of $f(x)=x^2\cdot x^8$.

This can be simplified to $f(x)=x^2\cdot x^8=x^{10}$. From this point you can quickly take the derivative using the rule $f'(x^n)=nx^{n-1}$. Therefore, $f'(x)=10x^9$.

Exercise 5

Find the derivative of $f(x)=(x^2-3)(x^3+1)$.

This is a product of two terms, which we can easily take the derivative of using the product rule. Recall that the product rule says for any two functions

$f(x)$ and $g(x)$

$(f \cdot g)'(x) = f'(x) \cdot g(x) + f(x) \cdot g'(x)$

This gives us:

$f'(x) = 2x(x^3 + 1) + (x^2 - 3)(3x^2)$ ←Apply the product rule.

$f'(x) = 2x^4 + 2x + 3x^4 - 9x^2$ ←Distribute.

$f'(x) = 5x^4 - 9x^2 + 2x$ ←Combine like terms and rearrange.

Exercise 6

Find the derivative of $f(x) = x^{-3}(3x^2 - 5x)$.

Again, you can use the product rule to find this derivative.

$(f \cdot g)'(x) = f'(x) \cdot g(x) + f(x) \cdot g'(x)$

This gives us:

$f'(x) = -3x^{-4}(3x^2 - 5x) + x^{-3}(6x - 5)$ ←Apply the product rule.

$f'(x) = \frac{-3(3x^2 - 5x)}{x^4} + \frac{6x - 5}{x^3}$ ←Rule of negative exponent.

You could simplify this further, but it is not necessary.

Exercise 7

Find the derivative of $f(x) = \frac{x^2}{2x}$.

Recall that the quotient rule says that for any two functions: $f(x)$ and $g(x)$, $\left(\frac{f}{g}\right)'(x) = \frac{f'(x) \cdot g(x) - f(x) \cdot g'(x)}{(g(x))^2}$

Let's substitute in our values to find the derivative.

$f'(x) = \frac{2x \cdot 2x - x^2 \cdot 2}{(2x)^2}$ ←Apply the quotient rule.

$f'(x) = \frac{4x^2 - 2x^2}{4x^2}$ ←Simplify.

$f'(x) = \frac{2x^2}{4x^2}$ ←Combine like terms.

$f'(x) = \frac{1}{2}$ ←Simplify.

Exercise 8

Find the derivative of $f(x) = e^{2x^3}$.

Recall from the quick reference sheet that $f'\left(e^{g(x)}\right) = g'(x)e^{g(x)}$. In this problem, $g(x) = 2x^3$. In order to complete this problem we will need to find $g'(x)$, which is $g'(x) = 6x^2$. If we apply the formula for the derivative to this problem, we get $f'(x) = 6x^2e^{2x^3}$.

Exercise 9

Find the derivative of $f(x) = \sin x \cos x$.

This is an application of the product rule as we're multiplying $\sin x \cos x$. You will also need the derivatives of the two trig functions sin *x* and cos *x*. You can find these on the quick reference sheet. They are:

$f'(\sin x) = \cos x$ and $f'(\cos x) = -\sin x$.

Recall that the product rule says $(f \cdot g)'(x) = f'(x) \cdot g(x) + f(x) \cdot g'(x)$.

For this problem:

$f'(x) = \cos x \cos x + \sin x(-\sin x)$ ←Apply the product rule.

$f'(x) = \cos^2 x - \sin^2 x$ ←Multiply.

TRY IT!

1. Find the derivative of the function $f(x) = 5x^2 + 3x - 1$ at the number 2.

2. Let $f(x) = \sqrt{x}$

   a. Find $f'(a)$.

   b. Find $f'(1)$, $f'(4)$, $f'(9)$

Answers

1. According to the definition of a derivative, with *a*=2, we have:

$$f'(2) = \lim_{h\to 0} \frac{f(2+h) - f(2)}{h} = \lim_{h\to 0} \frac{\left[5(2+h)^2 + 3(2+h) - 1\right] - \left[5(2)^2 + 3(2) - 1\right]}{h}$$

$$= \lim_{h\to 0} \frac{20 + 20h + 5h^2 + 6 + 3h - 1 - 25}{h}$$

$$= \lim_{h\to 0} \frac{23h + 5h^2}{h} = \lim_{h\to 0}(23 + 5h) = 23$$

2.

a) We use the definition of the derivative at *a*:

$$f'(x) = \lim_{h \to 0} \frac{f(x+h) - f(x)}{h}$$

$$= \lim_{h \to 0} \frac{\sqrt{a+h} - \sqrt{a}}{h}$$

$$= \lim_{h \to 0} \frac{\sqrt{a+h} - \sqrt{a}}{h} \cdot \frac{\sqrt{a+h} + \sqrt{a}}{\sqrt{a+h} + \sqrt{a}}$$

$$= \lim_{h \to 0} \frac{(a+h) - a}{h(\sqrt{a+h} + \sqrt{a})}$$

$$= \lim_{h \to 0} \frac{h}{h(\sqrt{a+h} + \sqrt{a})}$$

$$= \lim_{h \to 0} \frac{1}{\sqrt{a+h} + \sqrt{a}}$$

$$= \frac{1}{\sqrt{a} + \sqrt{a}}$$

$$= \frac{1}{2\sqrt{a}}$$

b) Substituting a=1, a=4, and a=9 into the result of part (a), we get

$f'(1) = \frac{1}{2\sqrt{1}} = \frac{1}{2}$ and $f'(4) = \frac{1}{2\sqrt{4}} = \frac{1}{4}$ and $f'(9) = \frac{1}{2\sqrt{9}} = \frac{1}{6}$

Example 5

Find the value of the definite integral $\int_0^4 x^2 dx$.

The integral often is called the **anti-derivative**. There are two types of integrals: the general integral and the definite integral. In the general integral, you must add an unknown constant to the solution since you have no numbers at which to evaluate the integral. However, in this example, we are told that we are to evaluate the integral from 0 to 4 so we will be able to find the actual value of this integral. Again, we have several common formulas for integrals on our reference sheet. If you check our reference sheet, you'll see that the value of $\int x^n dx = \frac{1}{n+1} x^{n+1} + c, n \neq 1$. If we apply that to this

example we get $\int x^2 dx = \frac{1}{2-1} x^{2+1} = \frac{1}{3} x^3$ which also can be written as $\frac{x^3}{3}$.
If this were a general integral, we would have to write the answer as $\frac{x^3}{3} + C$, however, our problem was a definite integral, so we can evaluate $\left.\frac{x^3}{3}\right]_0^4 = \frac{4^3}{3} - \frac{0^3}{3} = \frac{64}{3} - 0 = \frac{64}{3}$.

Exercise 1

Find the general integral $\int x^{-6} dx$.

Recall from the quick reference sheet that $\int x^n dx = \frac{1}{n+1} x^{n+1} + c, n \neq 1$.
If we apply this to this problem then:

$\int x^{-6} dx = \frac{1}{-6+1} x^{-6+1} + c$ ←Apply the formula for the integral.

$= -\frac{1}{5} x^{-5} + c$ ←Simplify.

Exercise 2

Find the value of the definite integral $\int_1^{10} x dx$.

Recall from the quick reference sheet that $\int x^n dx = \frac{1}{n+1} x^{n+1} + c, n \neq 1$.
If we apply this to find the general integral, then we get:

$$\int_1^{10} x dx = \frac{1}{1+1} x^{1+1} dx = \frac{1}{2} x^2.$$

Now, we'll evaluate this value for the integral from 1 to 10.

$$\left.\frac{1}{2} x^2\right]_1^{10} = \frac{1}{2} \cdot 10^2 - \frac{1}{2} \cdot 1^2 = \frac{1}{2} \cdot 100 - \frac{1}{2} \cdot 1 = 50 - \frac{1}{2} = 99/2$$

Exercise 3

Find the general integral $\int x^{-1}dx$.

If we apply the formula $\int x^n dx = \frac{1}{n+1}x^{n+1} + c, n \neq 1$, you get a 0 in the denominator of the fraction. You might think this means the integral is undefined, however, it's important to recall that x^{-1} can be written as $\frac{1}{x}$. Therefore, we can rewrite the problem as $\int \frac{1}{x}dx$. If you take a look at the quick reference sheet, you'll see that $\int \frac{1}{x}dx = \ln|x| + c$.

Calculus Quick Reference Sheet

Definition of the Derivative

$$f'(x) = \lim_{h\to 0}\frac{f(x+h)-f(x)}{h}$$

Common Derivatives
(*c* is any constant)

$f'(c) = 0$

$f'(x) = 1$

$f'(cx) = c$

$f'(x^n) = nx^{n-1}$

$f'(cx^n) = ncx^{n\;1}$

Derivatives of Trig Functions

$f'(\sin x) = \cos x$

$f'(\cos x) = -\sin x$

$f'(\tan x) = \sec^2 x$

$f'(\sec x) = \sec x \tan x$

$f'(\csc x) = -\csc x \cot x$

$f'(\cot x) = -\csc^2 x$

Derivatives of Logarithmic and Exponential Functions

$f'(c^x) = c^x \ln(c)$

$f'(e^x) = e^x$

$f'(\ln(x)) = \frac{1}{x}, x > 0$

$f'\left(e^{g(x)}\right) = g'(x)e^{g(x)}$

$f'(\ln|x|) = \frac{1}{x}, x \neq 0$

$f'(\ln g(x)) = \frac{g'(x)}{g(x)}$

$f'(\log_c(x)) = \frac{1}{x \ln c}, x > 0$

Common Rules & Formulas for Derivatives

Product Rule

For any two functions:

$f(x)$ and $g(x)$ $(f \cdot g)'(x) = f'(x) \cdot g(x) + f(x) \cdot g'(x)$

Quotient Rule

For any two functions $f(x)$ and $g(x)$ says

$$\left(\frac{f}{g}\right)'(x) = \frac{f'(x) \cdot g(x) - f(x) \cdot g'(x)}{(g(x))^2}$$

Common Integrals

$\int x^n dx = \frac{1}{n+1} x^{n+1} + c, n \neq 1$

$\int \frac{1}{x} dx = \ln|x| + c$

Integrals of Trig Functions

$\int \sin x dx = -\cos x + c$

$\int \csc x dx = \ln|\csc x - \cot x| + c$

$\int \cos x dx = \sin x + c$

$\int \sec x dx = \ln|\sec x + \tan x| + c$

$\int \tan x dx = \ln|\sec x| + c$

Terms You Should Know

Derivative - The derivative is an attempt to describe the motion of a moving object. You will hear the derivative called "the tangent line to the curve."

Integral - The Integral often is thought of as the anti-derivative.

Limits - The limit of a function describes the behaviors of a function as it approaches a specified value.

Quotient Rule - The quotient rule for any two functions $f(x)$ and $g(x)$ says $\left(\frac{f}{g}\right)'(x) = \frac{f'(x)\cdot g(x) - f(x)\cdot g'(x)}{(g(x))^2}$

Need a math tutor? Consider using one of TutaPoint.com's U.S.-based experts at www.tutapoint.com

Statistics

The study of statistics involves work in describing a situation that is occurring, as well as using this description to try to predict an outcome. The actual word "statistics" can be used to describe two different things. The first is an actual set of statistics about a person, event, or situation. For example, an athlete's statistics are the various numbers, such as points scored per game or amount of miles run, that describe the athlete's performance.

The other meaning of the word "statistics" is to describe the study of the branch of mathematics that involves the methods for planning experiments and simulations, gathering data based on these experiments and simulations, interpreting this data, and using the interpretations to draw conclusions for the data.

Statistics study often goes hand in hand with the study of **probability**, which is the field primarily involved with making predictions based on the statistics gathered from a sample in an experiment or from a population. A **sample** is a group of people or objects that are part of a larger population of all the people or objects that could be studied to answer a question. The process of obtaining a sample—selecting a smaller group of people or objects to study to make predictions about them—is an important part of statistics. Selecting an appropriate sample leads to more reliable data, which is used to draw conclusions about the larger population. Samples are used because it is not always possible to gather data from an entire population.

Statistics is about interpreting and analyzing data. In general, statistics is not an exact practice. While there are guidelines and rules to follow and calculations that must be completed accurately, the actual interpretation and drawing of conclusions is highly subjective. Individuals completing statistical research must be careful to avoid bias—the process of selecting samples or interpreting statistics in a way that leads a person to a desired viewpoint rather than an accurate viewpoint.

Example 1: Mean, Median, and Mode

Find the mean, median, and mode of the quiz scores 8, 4, 3, 10, 6, 9, 9, 7, 6, 10, 9.

Taken together, the words mean, median, and mode are called **measures of central tendency**. (**THINK:** The center or middle of a data set.) The mean of a set of data is the arithmetic average of all the values in the sample. The median of a set of data is the middle value when the scores are written in order from least to greatest. If there is an even number of values, the median is the average of the two middle numbers. The mode is the number that occurs the most frequently in a set of data. A set of data can have no mode (if the set contains every element exactly the same number of times) or it can be bimodal or multimodal, where more than one number occurs with the same frequency.

To find the mean of a sample, which is typically denoted by the symbol $\bar{x}$, add all the numbers in the set and divide by the number of data elements in the set. The number of data elements in the set is typically denoted by the letter *n*. This is also called the sample size. A summation sign $\sum$ is typically used to denote the concept of adding up all the numbers in the data set. So, to find the mean, do the following:

$$\text{Mean } (\bar{x}) = \frac{\sum x}{n} = \frac{8+4+3+10+6+9+9+7+6+10+9}{11} = \frac{81}{11} = 7.36$$

To find the median, write all the numbers in order from the least to the greatest and then identify the middle number (or in case of an even number of values, the middle values which will be averaged). The median typically is represented by the symbol $\tilde{x}$. To find the median, write the numbers in order from least to greatest:

3, 4, 6, 6, 7, 8, 9, 9, 9, 10, 10

Now identify the middle number or numbers in the set:

3, 4, 6, 6, 7, (8), 9, 9, 9, 10, 10

So, in this example, the median is 8.

Looking at the numbers when arranged in order from least to greatest is also the easiest way to find the mode. In this example, it is easy to see that the number 9 occurs more times than any other value in the list so the mode is 9.

Exercise 1

Find the **five-number summary** for the set of data shown below:

18 19 24 25 25 33 33 34 34 37 37 40 42 46 49 73

The five-number summary contains the minimum and maximum values of the data set, the median, the first quartile, and the third quartile values. So, in this case, the minimum is 18 and the maximum is 73. The median is the middle number or the average of the middle numbers (if the number of values is even). In this case the median will be the average of two values as shown below:

18 19 24 25 25 33 33 (34 34) 37 37 40 42 46 49 73

Since the two numbers are the same, their average is still 34 so the median is 34.

To find the first quartile, find the median of the data elements that are located to the left of the overall median. To find the third quartile, find the median of the data elements that are located to the right of the overall median.

18 19 24 2(5 25 3)3 33 34 | 34 37 3(7 40 4)2 46 49 73

Median

$$Q_1 = \frac{25+25}{2} = 25 \qquad Q_3 = \frac{40+42}{2} = 41$$

So, the five-number summary would be 18, 25, 34, 41, 73.

Exercise 2

Find the range of the set of data 27, 18, 19, 35, 22, 20, 14, 36, 25, 25, 42, 9, 17.

The **range** is the difference of the high and the low numbers in a set of data. You can rearrange the data in order from least to greatest if you like (or if you need to find the five-number summary as a later step in the problem). In a problem where you are only interested in the range, you can simply subtract the low number from the high number. In this case, the low number is 9 and the high number is 42. So the range = 42 – 9 = 33.

Exercise 3

Determine if there are any **outliers** in the data set from Exercise 1:

18 19 24 25 25 33 33 34 34 37 37 40 42 46 49 73

An outlier is defined as a piece of data in the sample that doesn't fit with the overall pattern of the data. This means the element of the data is unusually high or unusually low when compared to the other data elements. Determining whether something is an outlier is done by finding the Interquartile Range (IQR), multiplying this by 1.5, and subtracting it from the lowest value of the data set and adding it to the highest value of the data set. The IQR is found by subtracting Quartile 1 (Q_1) from Quartile 3 (Q_3). In this case that would be 41 – 25 = 16. We can now use this number to determine if any of the values are outliers. To do this, we need to do the following calculations:

Upper value for outliers: $Q_3 + (1.5 \cdot IQR) = 41 + (1.5 \cdot 16) = 41 + 24 = 65$

Lower value for outliers: $Q_1 - (1.5 \cdot IQR) = 25 - (1.5 \cdot 16) = 25 - 24 = 1$

This means if a value in the data set lies outside of the values from 1 to 65, then it is an outlier. In this case, 73 would be considered an outlier since it is outside of this set of values.

TRY IT!

1. Find the five-number summary for the data set shown: 2, 7, 11, 2, 1, 15, 8, 9, 3
2. Find the range of the set of data: 99, 75, 62, 98, 82, 67
3. Determine if there are any outliers in the data set used in Question 1.

Answers

1. The five-number summary contains the minimum and maximum values of the data set, the median, the first quartile, and the third quartile values. In this case, the minimum value is 1 and the maximum is 15. The median is the middle number of the average of the middle numbers. Let's begin by listing the numbers from least to greatest:

|

1, 2, 2, 3, 7, 8, 9, 11,15

In this case, the median or middle number is 7.

The first quartile is the median of the lower numbers and the third quartile is the median of the higher numbers. So, Q1=2 and Q3=10.

So, the five-number summary would be: 1, 2, 7, 10, 15

THINK: minimum, Q1, median, Q2, maximum.

2. The range is 99 - 62 or 37.

3. Using the formula discussed in exercise 3:

Upper value for outliers:

$$Q_3 + (1.5 \cdot IQR) = 10 + (1.5(10-2)) = 10 + 1.5(8) = 10 + 12 = 22$$

Lower value for outliers:

$$Q_1 - (1.5 \cdot IQR) = 2 - (1.5(10-2)) = 2 - 1.5(8) = 2 - 12 = -10$$

This means if a value in the data set is less than -10 or greater than 22, then it is an outlier. In this case, there are no outliers.

Example 2: Standard Deviation

The **standard deviation**, *s*, of a set of data, is the measure of how spread out the data is. Find the standard deviation of the following data set:

42, 100, 77, 54, 93, 85, 67, 77, 62, 58

Most modern statistics courses emphasize the use of graphing calculators or statistics software, both of which help to perform statistical calculations like this one. You will need to follow the instructions for your particular calculator or computer software to compute the standard deviation. We will demonstrate the mathematics that the calculator uses to compute the standard deviation.

Standard deviation is found by using the formula $s=\sqrt{\frac{\sum(x-\bar{x})^2}{n-1}}$. This says to find the value of the mean ($\bar{x}$) and subtract it from each of the elements in the set. Then square these individual results and add up the results. Divide by one less than the total number of elements in the set and then take the square root. You can see each of these steps demonstrated below:

Find the mean $\bar{x}$ =

$$\frac{42+100+77+54+93+85+67+77+62+58}{10}=\frac{715}{10}=71.5$$

Subtract the mean $\bar{x}$ from each of the elements in the set and then square each of these values (you may find it easier to list the data elements in a table like the one shown here to keep your work organized).

X	X - $\bar{x}$	$(x-\bar{x})^2$
42	-29.5	870.25
100	28.5	812.25
77	5.5	30.25
54	-17.5	306.25
93	21.5	462.25
85	13.5	182.25

67	-4.5	20.25
77	5.5	30.25
62	-9.5	90.25
58	-13.5	182.25

Add up all the values of $(x-\overline{x})^2$. In this example, the sum of these values is 2986.5.

Now, find n – 1. In this case, there are 10 elements in the set so n – 1 = 10 – 1 = 9.

Now divide these two numbers and take their square root.

$$s=\sqrt{\frac{2986.5}{9}}=\sqrt{331.83}=18.22$$

Exercise 1

Find the variance of the set of data 8, 9, 7, 10, 5, 7.

Variance is the square of standard deviation. As you'll recall, standard deviation can be found by evaluating the formula $s=\sqrt{\frac{\sum(x-\overline{x})^2}{n-1}}$. Since variance is equal to the square of standard deviation, the variance can be found by evaluating the formula $s^2=\frac{\sum(x-\overline{x})^2}{n-1}$. Follow the same process followed in Example 2 to find the standard deviation, but do not take the square root in the last step.

1. Find the mean $\overline{x}=\frac{8+9+7+10+5+7}{6}=\frac{46}{6}=7.7$

2. Subtract the mean $\overline{x}$ from each of the elements in the set and then square each of these values (you may find it easier to list the data elements in a table like the one shown here to keep your work organized).

X	$x - \overline{x}$	$(x-\overline{x})^2$
8	0.3	0.09
9	1.3	1.69
7	-0.7	0.49
10	2.3	5.29
5	-2.7	7.29
7	-0.7	0.49

3. Add up all the values of $(x-\overline{x})^2$. In this example, the sum of these values is 15.34.

4. Now, find $n - 1$. In this case, there are six elements in the set so

$n - 1 = 6 - 1 = 5$.

5. Now divide these two numbers:

$$s^2 = \frac{15.34}{5} = 3.068$$

It's important to note that in statistics, each variable has a specific meaning, so while taking the square root of this number will give you s, we are not necessarily interested in that. Therefore you do not need to follow the protocols for algebra in each problem and you don't need to take the square root of both sides to answer this question.

Example 3: Probability

On a quiz consisting of three multiple choice questions (where only one selection from the list of possible answers given is correct), a student who was not prepared must guess at each answer. Find the probability that the student will earn a passing grade on the quiz (2 or 3 correct responses).

Probability is defined as the number of ways a desired event can occur divided by the total number of possible events. You often see this written as:

$$P(A) = \frac{\text{number of ways an event can occur}}{\text{total number of possible events}}.$$

So, in order to work this problem, we must find out these two values. Let's consider this situation for just a minute. There are two possibilities that can occur on each question: The student can either get the question right (R) or wrong (W).

So, the possible situations could result:

Question	Case 1	Case 2	Case 3	Case 4	Case 5	Case 6	Case 7	Case 8
1	W	R	W	W	R	R	W	R
2	W	W	R	W	R	W	R	R
3	W	W	W	R	W	R	R	R

So, there are 8 total possible events. Now, we need to count the number of cases in which you get the desired result, which is two or more right answers. In the table, you can see that this occurs in cases 5, 6, 7, and 8. To find the probability, write the number of times you get the desired result (4) over the total number of possible events (8).

$$P(A) = \frac{4}{8} = 0.5$$

This means there is a 50% chance of getting a passing grade.

THINK: Question 1-Right or Wrong-2 options

Question 2-Right or Wrong-2 options

Question 3-Right or Wrong-2 options

So, (2)(2)(2)=8

Example 4: Normal Distribution

The height of adult women in the United States is normally distributed with a mean of 65 inches and a standard deviation of 1.2". Find the probability that a woman's height is between 63.8" and 66.2".

A normal distribution can be used to describe a variety of naturally occurring phenomena (such as human characteristics like height in this example), industrial manufacturing, science, and behavioral characteristics. The **normal distribution** is represented by a bell-shaped curve—often simply called the bell curve. The bell curve looks similar to the graphic shown below:

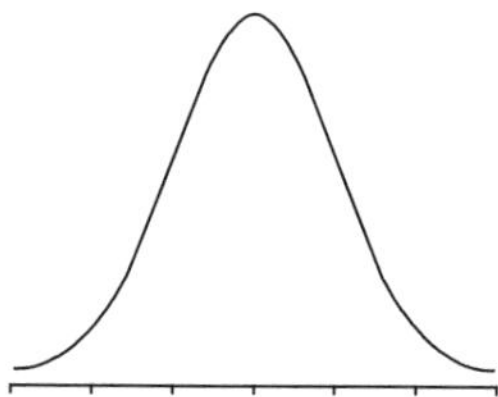

The normal distribution can be relatively tall and not wide spread (like the one shown above). If that is the case, the values of the data elements tend to be more closely centered around the mean, which means the standard deviation is smaller. If the normal distribution is shorter and wider, then the standard deviation will be larger, which means the differences between the individual values will vary more from one to the next.

In every normal distribution, data behaves in a predictable pattern. Approximately 68% of the data is centered around the mean within 1 standard deviation. Approximately 95% of the data is located within 2 standard deviations of the mean, and approximately 99% of the data is located within 3 standard deviations of the mean. This is summarized in the graphs below.

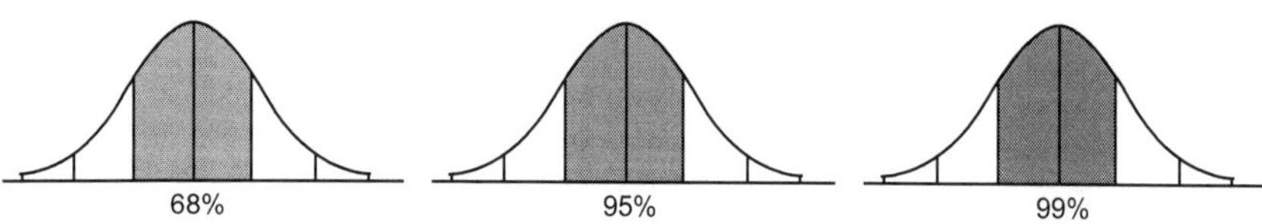

In this example, we simply need to determine how many standard deviations from the mean the range we have been given is, and we can then determine the percentage of results that fit this description.

The distribution we are investigating has a mean of 65 inches and a standard deviation of 1.2". This tells us that values located 1 standard deviation from the mean would be found by taking 65 ± 1.2.

To find the lower value, subtract 1.2 from 65, which gives 63.8. To find the upper value add 1.2 to 65, which gives 66.2. Since this is the range we are looking for in the problem, we know that this range lies one standard deviation from the mean. This means 68% of the population will fall within this range.

Exercise 1

Scores on the SAT math exam of students at Galaxy High School are normally distributed with a mean of 580 and a standard deviation of 120 points. Into what score range can you expect 95% of the students to fall?

Recall that in a normal distribution, 95% of the data lies within 2 standard deviations of the mean. To solve this problem we need to calculate the mean $\pm$ 2 standard deviations. To do this we would compute: $580 \pm 2(120) = 580 \pm 240 = 340$ and 820.

Exercise 2

There are 980 students in the class at Galaxy High School. What score would you expect the top 1% of this class to earn on the math portion of the SAT if the distribution in Exercise 1 holds true?

The top 1% of the class will score just slightly above and below the third standard deviation above the mean (remember that 99% of the data falls within 3 standard deviations of the mean, so ½ of a percent lies above the mean and ½ of a percent lies below the mean; therefore the top 1% will be located very near the third standard deviation, but not exactly on it). To find the third standard deviation above the mean we calculate:

$580 + 3(120) = 580 + 360 = 940$. Therefore, we can expect the top 1% of the class to have scores somewhere around 940.

Example 5: Data Correlation

A set of data, when placed on a graph, looks like the graph shown below. Describe the **correlation** of this data.

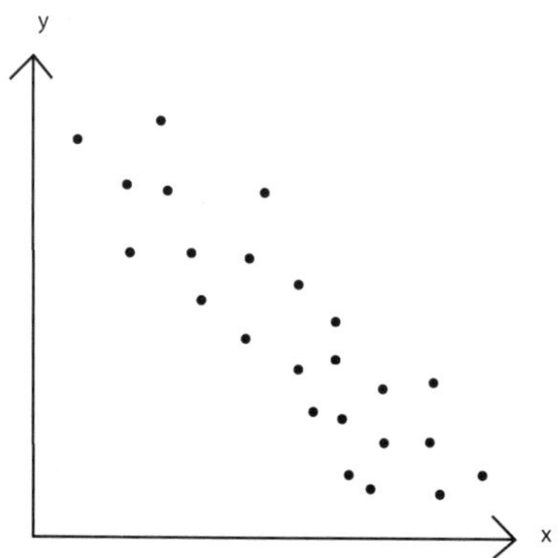

The graph shown here represents a set of data that is roughly correlated. If the data points were more spread out and did not exhibit any sort of pattern, it could be said that the data is not correlated. However, this data follows a pattern that appears to decrease from left to right. When data follows a pattern that moves down from left to right, the data is said to be negatively correlated. The more tightly grouped the data points are, the more highly correlated. These data points are relatively spread apart (a perfect correlation would lie in a completely straight line). Since these data points are relatively spread apart, this is said to be a low, negative correlation.

THINK: What would positive or no correlation look like? Positive correlation would increase from left to right. No correlation would not increase from left to right.

Exercise 1

Describe the correlation of the data shown below.

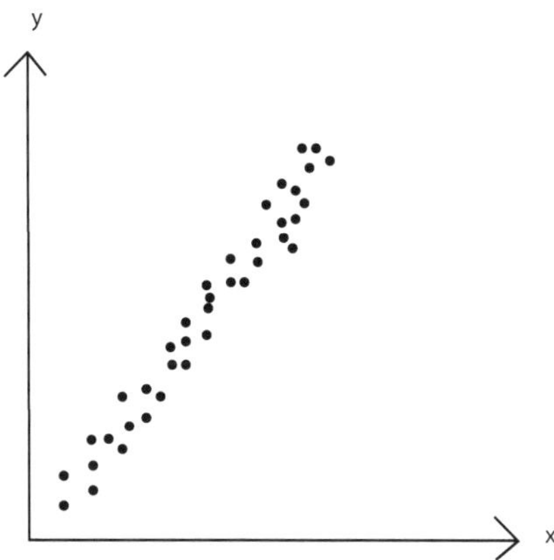

This data set is increasing from left to right, so the data is said to be positively correlated. Since the data points fall into a nearly perfect straight line, the data is said to be strongly correlated. To completely describe this data set we would say it has a strong, positive correlation.

Exercise 2

Describe the correlation of the data shown below.

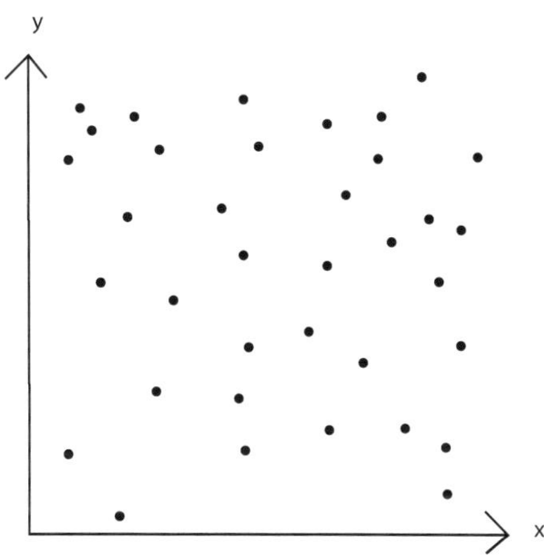

There does not appear to be any pattern in the data shown here. When there is no pattern or the pattern is inconsistent, we say that there is no correlation.

Exercise 3

There are five red balls, five blue balls, and two orange balls in a bag. What is the probability of drawing an orange ball out of the bag?

The probability can be found by applying the following:

$$P(A) = \frac{\text{number of ways a desired event can occur}}{\text{total number of possible events}}$$

In this case, there are two ways that an orange ball can be drawn out of the bag, so the numerator will be 2. There are 12 possible outcomes or events that can occur (each of the balls that could be drawn out of the bag). So,

$$P(A) = \frac{2}{12} = \frac{1}{6}$$

Exercise 4

A statistics teacher keeps detailed records of the results of the final exam he gives every year. He records the number of students who take the exam each year and their grades and adds this to his total. After giving this exam for five years, the distribution of his grades is shown in the following table. In the teacher's current class of 25 students, using this distribution, how many students should the teacher expect to earn a grade of A?

Letter Grade	Number of Students	Percent of Students
A	74	13 %
B	139	24 %
C	192	33 %
D	101	18 %
F	68	12 %

We can use this percentage distribution to make predictions about future classes, which is what we are being asked to do here. In the distribution shown, 13% of students earn a grade of A. If the distribution continues, we would expect 13% of the current class of 25 students to earn a grade of A. We can find out what number this is by taking 13% of 25. This is done by performing the following calculation: $0.13 \cdot 25 = 3.25$ students. Since we cannot have 3.25 students, we would say that 3 or 4 students can be expected to earn a grade of A on the exam.

Exercise 5

A bag of chocolate coated candies contains candy in five colors: purple, green, blue, red, and yellow. The manufacturer of these candies indicates that they place each color of candy in the bag using the distribution shown in the pie chart below. If this distribution is true and there are 18 red candies in the bag, how many candies in total are in the bag?

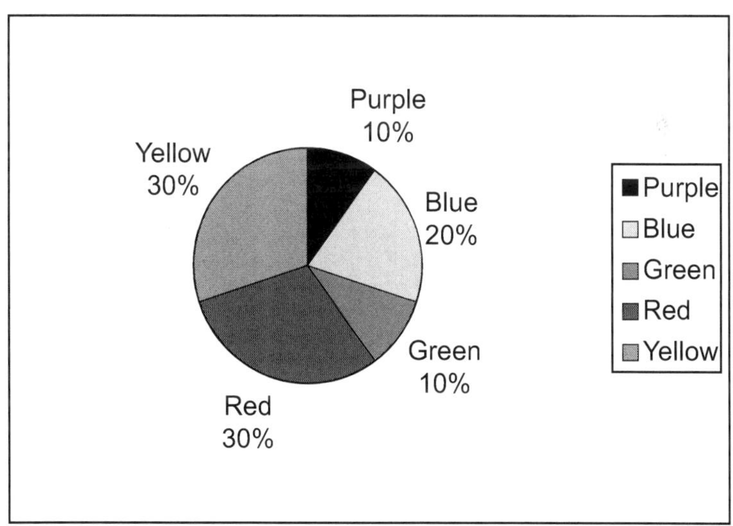

If there are 18 red candies in the bag, then that means that 30% of the bag's contents is 18. So, we want to know the answer to the following question to solve this problem: 18 is 30% of what number? Using a small bit of translation from words to symbols we can rewrite this problem as $18 = 0.30 \cdot x$. To solve this problem, divide both sides by 0.30 so $x = 60$. Based on this data, there are 60 candies in the bag.

Exercise 6

A high school basketball team scores the following number of points per game over a single season:

76	74	82	96	66	76	78	72	52	68
86	84	62	76	78	92	82	74	88	63

Construct a stem-and-leaf plot of this data.

A **stem-and-leaf plot** is a method for representing data so that it develops a somewhat graphical appearance. In this method of writing data, you write the first numbers of the data elements on one side of a line (stem) and then list the second numbers in a row to the right of the first number (leaf). To write the first element, we would write:

$$7|6$$

To add the second element to this we would write:

$$7 \mid 6 \quad 4$$

So, now we simply do this same step with all the elements:

$$\begin{array}{c|cccccccc} 5 & 2 \\ 6 & 6 & 8 & 2 \\ 7 & 6 & 4 & 6 & 8 & 2 & 6 & 8 & 4 \\ 8 & 2 & 6 & 4 & 2 & 8 \\ 9 & 6 & 2 \end{array}$$

We would complete the stem-and-leaf plot by re-writing each row in order from least to greatest as shown:

Stem	Leaf
5	2
6	2 6 8
7	2 4 4 6 6 6 8 8
8	2 2 4 6 8
9	2 6

Key: 6|2 = 62

Exercise 7

A married couple has decided they want to have three children. Describe a simulation you could run to tell them the likelihood that two of their children will be boys and one will be a girl.

A **simulation** is a way of representing a situation that would be difficult to complete in real life but can be generated in a laboratory or experimental environment. In this case, we know that the probability of having a boy and the probability of having a girl is roughly equal at 50% each. So, there are many ways we could simulate having three children. We could toss a coin three times and record the result. If we let heads mean that the couple had a girl and tails mean they had a boy then we could simply count up the number of times we got two tails and one head and divide by the total number of times we ran the simulation.

Another way of doing this is by rolling a six-sided dice. If you let the odd numbers (1, 3, and 5) represent having a boy and even numbers (2, 4, and 6) represent having a girl, then each time you roll the die three times, record the results and count up the number of times you got a simulation showing two boys (two odd numbers) and one girl (an even number).

You could also use a random number table or a random number generator on the computer or calculator. These programs generate a number between 0 and 1, so for this example you could say any value below 0.5 would represent a girl and any value greater than or equal to 0.5 would represent a boy. Again, record the results and count the number of times you got two boys and one girl.

If you conduct the simulation five times (either by tossing a coin, rolling a die, or getting a random number), you'll get a probability that is relatively low. If you continue running the simulation over and over—which is where computers and graphing calculators come in handy—you'll begin to see that regardless of how many times you run the simulation the probability approaches the same value. This value, which is found after running the simulation many, many times, is close to the true probability.

It's important to note that even if you run the experiment 1,000,000 times or more, there is no way to guarantee to the couple what the outcome will be when they have children.

Exercise 10

Describe the differences in the two normal distributions pictured below.

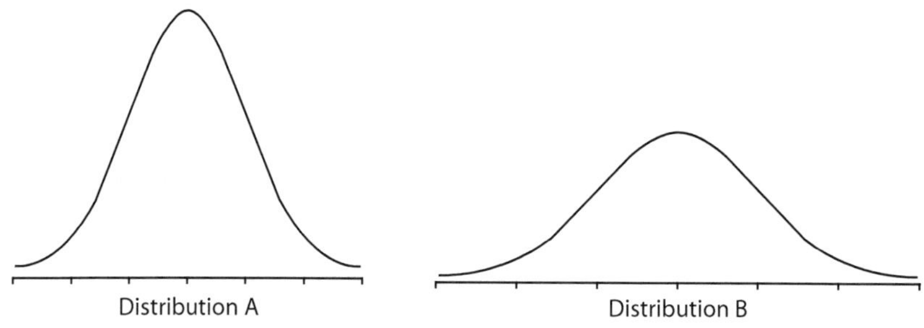

Recall that the picture of the distribution helps us to understand how closely the data points are gathered around the mean. Distribution A is going to have a smaller standard deviation than Distribution B. Also, the range of values in Distribution A most likely will be smaller.

Exercise 11

The number of hours you spend studying for a statistics test is positively correlated with the score you will receive on the exam. If Heather spends 3 hours studying for the exam and Jennifer spends 4.5 hours studying for the exam, whose score would you expect to be higher?

Negative correlation means that as one value increases the other decreases and vice-versa. Positive correlation means that as one value increases (or decreases) the other value also increases (or decreases). So, if the number of hours spent studying increases, you would expect the test score to increase as well, since they are positively correlated. Therefore, we would expect Jennifer to receive the better grade on the exam.

Terms You Should Know

Correlation - A correlation is a description of the association between coordinates.

Five-Number Summary - The five-number summary contains the minimum and maximum values of the data set, the median, the first quartile, and the third quartile value.

Measures of Central Tendency - The center or middle of a data set is measured using the measures of central tendency (mean, median, mode). The mean of a set of data is the arithmetic average of all the values in the sample. The median of a set of data is the middle value when the scores are written in order from least to greatest. If there is an even number of values, the median is the average of the two middle numbers. The mode is the number that occurs the most frequently in a set of data.

Outlier - An outlier is a piece of data in a sample that doesn't fit with the overall pattern of the data.

Probability - Probability is the field primarily involved with making predictions based on the statistics gathered from a sample in an experiment or from a population.

Range - The range is the difference of the high and the low numbers in a set of data.

Sample - A sample is a group of people or objects that are part of a larger population of all the people or objects that could be studied to answer a question.

Standard Deviation - The standard deviation, s, of a set of data, is the measure of how spread out the data is.

Stem and Leaf Plot - A stem-and-leaf plot is a method for representing data so that it develops a somewhat graphical appearance.

Variance - Variance is the square of standard deviation.

At TutaPoint we strive to make your learning experience as efficient and effective as possible. Working with curriculum professionals from across the country we have developed a series of pre-recorded mini lessons of the most common core concepts students have difficulty mastering. Lessons can be found on our website, www.TutaPoint.com, and can be downloaded to your computer or portable device. Additionally, our tutors use results from our online quizzes to help them pinpoint students' strengths and weaknesses, resulting in extremely productive sessions. Learn more at www. TutaPoint.com.

Mini Lesson Topic Overview and Index

Essential Math Skills/Pre-Algebra Topics

1. Identifying fractions from a shaded region
2. Finding prime factorizations
3. Reducing fractions to lowest terms
4. Place value of decimals and whole numbers
5. Finding least common denominators
6. Convert decimals to fractions
7. Convert fractions to decimals
8. Convert decimals to percents and percents to decimals
9. Convert percents to fractions
10. Comparing numbers using less than, greater than, less than or equal to, and greater than or equal to
11. Ordering positive, real numbers on a number line

12. Improper Fractions & Mixed Numbers
13. Adding and subtracting fractions less than 1
14. Adding and subtracting fractions greater than 1
15. Multiplying and dividing fractions
16. Adding and subtracting decimals
17. Multiplying decimals
18. Dividing decimals
19. Adding and subtracting integers
20. Multiplying and dividing integers
21. Evaluating exponents
22. Order of Operations

Algebra I

1. Basic graphing skills
2. Solving one-step equations using number sense
3. Solving one-step equations algebraically
4. Solving two-step equations
5. Solving multi-step equations
6. Applications of solving equations
7. Finding slope from two points
8. Finding x- and y- intercept
9. Finding the equation of a line given the slope and a point on the line
10. Finding the equation of a line given two points on the line
11. Relationships between equations, tables of values, and graphs

12. Graphing linear equations using the slope and intercepts
13. Applications of linear equations
14. Direct variation and its applications
15. Indirect variation and its applications
16. Applications of variation
17. Simplification of simple rational expressions
18. Scientific Notation
19. Identifying Greatest Common Factor
20. Factoring out the Greatest Common Factor
21. Factoring the difference of squares
22. Factoring trinomials 1
23. Solving equations using factoring

Algebra II

1. Find the solution of systems of linear equations by graphing
2. Find the solution of systems of linear equations by substitution
3. Find the solution of systems of linear equations by elimination
4. Find the solution of systems of linear equations using matrices
5. Identifying a parabola
6. Identifying a circle
7. Identifying an ellipse
8. Identifying a hyperbola
9. Identifying conic sections
10. Evaluating logarithms
11. Operations with logarithms

12. Applications involving exponential/logarithmic growth
13. Solving equations involving absolute value
14. Solving radical equations
15. Solving radical equations
16. Solving rational equations
17. Domain and range of functions
18. Evaluating composition of functions
19. Finding rational roots

Geometry

1. Finding the distance between two points
2. Finding the midpoint of a line segment
3. Angle relationships – acute, obtuse, and right angles
4. Angle relationships – angle bisectors
5. Line relationships – parallel and perpendicular
6. Alternate angles
7. Consecutive angles
8. Applications of linear and angular relationships
9. Triangles and their relationships
10. Polygons
11. Similarity
12. Properties of shapes in coordinate plane
13. Special points and circular relationships
14. Angular relationships with circles
15. Reflection and transformation

16. Area
17. Pythagorean Theorem
18. Volume
19. Geometric Probability

Trigonometry

1. Radians and degrees
2. 30-60-90 right triangle
3. 45-45-90 right triangle
4. The Unit Circle
5. Sine
6. Cosine
7. Tangent
8. Cosecant
9. Secant
10. Cotangent
11. Trig Identities
12. Graphing Sine
13. Graphing Cosine
14. Graphing Tangent
15. Solving trigonometric equations

Pre-Calculus

1. Rational Root Theorem
2. Solving rational equations

3. Applications of exponential and logarithmic growth
4. Domain and range of rational equations
5. Decompose fractions
6. Evaluating functions
7. Composition of functions
8. Operations with vectors
9. Properties of vectors

Calculus

1. Evaluating limits
2. The definition of a derivative
3. Finding derivatives
4. Introduction to integration
5. Evaluating definite integrals

Statistics

1. Measures of central tendency
2. Five-number summary and box plots
3. Circle graphs
4. Correlation
5. The normal distribution
6. Probability

Math Made Simple
Scope and Sequence
Correlating with NCTM Standards
National Council of Teachers of Mathematics

TutaPoint.com has written “Math Made Simple using the principles and standards for school mathematics published by the National Council of Teachers of Mathematics (NCTM). These standards are the cornerstone of basic mathematic principles that ensure the highest quality of learning for students.

We at TutaPoint.com recognize the importance of consistent standards of learning and have correlated our supplemental text with the NCTM principles and standards. The next several pages provide a scope and sequence of our content to NCTM’s guidelines.

Number Operations

TutaPoint Mini Lesson Scope and Sequence

		TutaPoint Mini Lesson Scope and Sequence							
	Understand numbers, ways of representing numbers, relationships among numbers, and number systems				Understand meanings of operations and how they relate to one another			Compute fluently and make reasonable estimates	
	Develop a deeper understanding of very large and very small numbers and of various representations of them	Compare and contrast the properties of numbers and number systems, including rational and real numbers, and understand complex numbers as solutions to quadratic euqations that do not have real solutions	Understand vectors and matrices as systems that have some of the properties of the real-number system	Use number-theory arguments to justify relationships involving whole numbers	Judge the effects of such operations as multiplication, division, and computing powers and roots on the magnitudes of quantities	Develop an understanding of properties of, and representations for, the addition and multiplication of vectors and matricies	Develop an understanding of permutations and combinations as counting techniques	Develop fluency in operations with real numbers, vectors, and matrices, using mental computation or paper-and-pencil calculations for simple cases and technology for more complicated cases	Judge the reasonableness of numerical computations and their results
Pre -Algebra/ Algebra I	X	X	X	X	X		X	X	X
Algebra II	X	X	X		X	X		X	X
Geometry		X	X		X			X	X
Pre-calculus	X	X	X		X	X		X	X
Calculus	X	X			X			X	X
Statistics		X			X			X	X
Trigonometry		X	X		X			X	X

Algebra

	Understand patterns, relations, and functions						Represent and analyze mathematical situations and structures using algebraic symbols					Use mathematical models to represent and understand quantitative relationships			Analyze change in various contexts
	Generalize patterns using explicitly defined and recursively defined functions	Understand relations and functions and select, convert flexibly among, and use various representations for them	Analyze functions of one variable by investigating rates of change, intercepts, zeros, asymptotes, and local and global behavior	Understand and perform transformations such as arithmetically combining, composing, and inverting commonly used functions, using technology to perform such operations on more complicated symbolic expressions	Understand and compare the properties of classes of functions, including exponential, polynomial, rational, logarithmic, and periodic functions	Interpret representations of functions of two variables	Understand the meaning of equivalent forms of expressions, equations, inequalities, and relations	Write equivalent forms of equations, inequalities, and systems of equations and solve them with fluency - mentally or with paper and pencil in simple cases and using technology in all cases	Use symbolic algebra to represent and explain mathematical relationships	Use a variety of symbolic representations, including recursive and parametric equations, for functions and relations;	Judge the meaning, utility, and reasonableness of the results of symbol manipulations, including those carried out by technology	Identify essential quantitative relationships in a situation and determine the class or classes of functions that might model the relationships	Use symbolic expressions, including iterative and recursive forms, to represent relationships arising from various contexts	Draw reasonable conclusions about a situation being modeled	Approximate and interpret rates of change from graphical and numerical data
Algebra	X	X	X	X	X	X	X	X	X	X	X	X	X	X	X
Algebra II	X	X	X	X	X	X	X	X	X	X	X	X	X	X	X
Geometry							X	X	X	X	X	X	X	X	
Pre-Calculus	X	X	X	X	X	X	X	X	X	X	X	X	X	X	X
Calculus	X	X	X	X	X		X	X	X	X	X	X	X	X	X
Statistics							X	X	X	X	X	X	X	X	
Trigonometry		X		X		X	X	X	X	X	X	X	X	X	

Geometry

	Analyze chracteristics and properties of two- and three-dimensional geometric shapes and develop mathematical arguments about geometric relationships				Specify locations and describe spatial relationships using coordinate geometry and other representational systems		Apply transformations and use symmetry to analyze mathematical situations		Use visualization, spatial reasoning, and geometric modeling to solve problems				
	Analyze properties and determine attributes of two - and three-dimensional objects	Explore relationships (including congruence ad similarity) among classes of two- and three-dimensional geometric objects, make and test conjectures about them, and solve problems involving them	Establish the validity of geometric conjectures using deduction, prove theorems, and critique arguments made by others	Use trigonometric relationships to determine lengths and angle measures	Use Cartesian coordinates and other coordinate systems, such as navigational, polar, or spherical systems, to analyze geometric situations	Investigate conjectures and solve problems involving two- and three-dimensional objects represented with Cartesian coordinates	Understand and represent translations, reflections, rotations, and dilations of objects in the plane by using sketches, coordinates, vectors, function notation, and matrices	Use various representations to help understand the effects of simple transformations and their compositions.	Draw and construct representations of two- and three-dimensional geometric objects using a variety of tools	Visualize three-dimensional objects and spaces from different perspectives and analyze their cross sections	Use vertex-edge graphs to model and solve problems	Use geometric models to gain insights into, and answer questions in, other areas of mathematics	Use geometric ideas to solve problems in, and gain insights into, other disciplines and other areas of interest such as art and architecture.
Algebra					X	X			X			X	X
Algebra II					X	X	X	X	X	X		X	X
Geometry	X	X	X	X	X	X	X	X	X	X		X	X
Pre-calculus					X	X	X	X	X	X		X	X
Calculus					X								
Statistics					X	X			X			X	X
Trigonometry	X			X	X	X	X		X			X	X

Measurement

	Understand measurable attributes of objects and the units, systems, and processes of measurment	Apply appropriate techniques, tools, and formulas to determine measurements			
	Make decisions about units and scales that are appropriate for problem situations involving measurement	Analyze precision, accuracy, and approximate error in measurement situations	Understand and use formulas for the area, surface area, and volume of geometric figures, including cones, spheres, and cylinders	Apply informal concepts of successive approximation, upper and lower bounds, and limit in measurement situations	Use unit analysis to check measurement
Algebra	X				
Algebra II	X	X	X		
Geometry	X		X		
Pre-calculus	X			X	
Calculus				X	
Statistics	X	X			
Trigonometry	X				

Data Analysis & Probability

	Formulate questions that can be addressed with data and collect, organize, and display relevant data to answer them					Select and use appropriate statistical methods to analyze data					Develop and evaluate inferences and predictions that are based on data				Understand and apply basic concepts of probability			
	Understand the differences among various kinds of studies and which types of inferences can legitimately be drawn from each	Know the characteristics of well-designed studies, including the role of randomization in surveys and experiments	Understand the meaning of measurement data and categorical data, of univariate and bivariate data, and of the term variable	Understand histograms, parallel box plots, and scatterplots and use them to display data	Compute basic statistics and understand the distinction between a statistic and a parameter	For univariate measurement data, be able to display the distribution, describe its shape, and select and calculate summary statistics	For bivariate measurement data, be able to display a scatterplot, describe its shape, and determine regression coefficients, regression equations, and correlation coefficients using technological tools	Display and discuss bivariate data where at least one variable is categorical	Recognize how linear transformations of univariate data affect shape, center, and spread	Identify trends in bivariate data and find functions that model the data or transform the data so that they can be modeled	Use simulations to explore the variability of sample statistics from a known population and to construct sampling distributions	Understand how sample statistics reflect the values of population parameters and use sampling distributions as the basis for informal inference	Evaluate published reports that are based on data by examining the design of the study, the appropriateness of the data analysis, and the validity of conclusions	Understand how basic statistical techniques are used to monitor process characteristics in the workplace	Understand the concepts of sample space and probability distribution and construct sample spaces and distributions in simple cases	Use simulations to construct empirical probability distributions	Compute and interpret the expected value of random variables in simple cases	Understand the concepts of conditional probability and independent events
Algebra																		
Algebra II																		
Geometry																		
Pre-calculus																		
Calculus																		
Statistics	X		X	X	X	X	X		X	X	X	X		X	X	X	X	X
Trigonometry																		

Problem Solving

	Problem Solving			
	Build new mathematical knowledge through problem solving	Solve problems that arise in mathematics and in other contexts	Apply and adapt a variety of appropriate strategies to solve problems	Monitor and reflect on the process of mathematical problem solving
Algebra	X	X	X	
Algebra II	X	X	X	
Geometry	X	X	X	
Pre-calculus	X	X	X	
Calculus	X	X	X	
Statistics	X	X	X	
Trigonometry	X	X	X	

Reasoning & Proof

	Reasoning and Proof			
	Recognize reasoning and proof as fundamental aspects of mathematics	Make and investigate mathematical conjectures	Develop and evaluate mathematical arguments and proofs	Select and use various types of reasoning and methods of proof
Algebra	X	X		X
Algebra II	X	X		X
Geometry	X	X	X	X
Pre-calculus	X	X		X
Calculus	X	X		X
Statistics	X	X		X
Trigonometry	X	X		X

Communication

	Communication			
	Organize and consolidate their mathematical thinking through communication	Communicate their mathematical thinking coherently and clearly to peers, teachers, and others	Analyze and evaluate the mathematical thinking and strategies of others	Use the language of mathematics to express mathematical ideas precisely
Algebra	X	X		X
Algebra II	X	X		X
Geometry	X	X		X
Pre-calculus	X	X		X
Calculus	X	X		X
Statistics	X	X		X
Trigonometry	X	X		X

Connections

	Connections			
	Recognize and use connections among mathematical ideas	Understand how mathematical ideas interconnect and build on one another to produce a coherent whole	Recognize and apply mathematics in contexts outside of mathematics	
Algebra	X	X	X	
Algebra II	X	X	X	
Geometry	X	X	X	
Pre-calculus	X	X	X	
Calculus	X	X		
Statistics	X	X	X	
Trigonometry	X	X	X	

Representation

	Representation		
	Create and use representations to organize, record, and communicate mathematical ideas	Select, apply, and translate among mathematical representations to solve problems	Use representations to model and interpret physical, social, and mathematical phenomena